BOTANICAL RESEARCH AND PRACTICES

ECOLOGICAL RANGES AND ECOLOGICAL NICHES OF PLANT SPECIES IN THE MONSOON ZONE OF PACIFIC RUSSIA

Botanical Research and Practices

Additional books in this series can be found on Nova's website under the Series tab.

Additional E-books in this series can be found on Nova's website under the E-book tab.

BOTANICAL RESEARCH AND PRACTICES

ECOLOGICAL RANGES AND ECOLOGICAL NICHES OF PLANT SPECIES IN THE MONSOON ZONE OF PACIFIC RUSSIA

VITALY P. SELEDETS
AND
NINA S. PROBATOVA

Nova Science Publishers, Inc.
New York

Copyright © 2012 by Nova Science Publishers, Inc.

All rights reserved. No part of this book may be reproduced, stored in a retrieval system or transmitted in any form or by any means: electronic, electrostatic, magnetic, tape, mechanical photocopying, recording or otherwise without the written permission of the Publisher.

For permission to use material from this book please contact us:
Telephone 631-231-7269; Fax 631-231-8175
Web Site: http://www.novapublishers.com

NOTICE TO THE READER

The Publisher has taken reasonable care in the preparation of this book, but makes no expressed or implied warranty of any kind and assumes no responsibility for any errors or omissions. No liability is assumed for incidental or consequential damages in connection with or arising out of information contained in this book. The Publisher shall not be liable for any special, consequential, or exemplary damages resulting, in whole or in part, from the readers' use of, or reliance upon, this material. Any parts of this book based on government reports are so indicated and copyright is claimed for those parts to the extent applicable to compilations of such works.

Independent verification should be sought for any data, advice or recommendations contained in this book. In addition, no responsibility is assumed by the publisher for any injury and/or damage to persons or property arising from any methods, products, instructions, ideas or otherwise contained in this publication.

This publication is designed to provide accurate and authoritative information with regard to the subject matter covered herein. It is sold with the clear understanding that the Publisher is not engaged in rendering legal or any other professional services. If legal or any other expert assistance is required, the services of a competent person should be sought. FROM A DECLARATION OF PARTICIPANTS JOINTLY ADOPTED BY A COMMITTEE OF THE AMERICAN BAR ASSOCIATION AND A COMMITTEE OF PUBLISHERS.

Additional color graphics may be available in the e-book version of this book.

Library of Congress Cataloging-in-Publication Data

Ecological ranges and ecological niches of plant species in the monsoon zone of Pacific Russia / editors: Vitaly P. Seledets, Nina S. Probatova.
   p. cm.
 Includes bibliographical references and index.
 ISBN 978-1-62100-434-9 (softcover)
 1. Plant ecology--Russia (Federation)--Russian Far East. 2. Niche (Ecology) I. Seledets, V. P. II. Probatova, N. S. (Nina Sergeevna)
 QK375.E25 2012
 581.747--dc23
                              2011035650

Published by Nova Science Publishers, Inc. † New York

This book is dedicated to the 40-th Anniversary of the Pacific Institute of Geography and to 50-th Anniversary of the Institute of Biology & Soil Science, the Far East Branch of the Russian Academy of Sciences, in which this work was carried out.

CONTENTS

Preface		ix
Acknowledgments		xi
Introduction: The Monsoon Zone in the Russian Far East		xiii
Chapter 1	Ecology of the Area of Study	1
Chapter 2	The Family Poaceae as a Model Group	11
Chapter 3	Materials and Methods	15
Chapter 4	The Concept of Ecological Range	23
Chapter 5	Ecological Differentiation	37
Chapter 6	Ecological Ranges of Plant Species in Inner Asia and the Pacific Region	41
Chapter 7	Ecological Ranges of Invasive Species	51
Chapter 8	Ecological Niches	55
Conclusion		73
References		75
Appendix 1.		111
Appendix 2.		133
About the Authors		143
Index		147

Preface

The problem of biodiversity in the monsoon zone is connected to species adaptations, speciation, florogenesis, plant community formation, vegetation dynamics and population structure. The authors of this book applied the concept of ecological range (ER) of plant species to investigate adaptive strategies of plants in the Russian Far East (RFE) monsoon zone. The ER is a part of multidimensional ecological space (MES), and part of the ecological niche. Every species has its own ER and ecological niche, and they change in different parts of the geographical range of the species.

ACKNOWLEDGMENTS

We wish to thank Dr. Elena Volynets, for her constant assistance during the preparation of this book, and Elvira Rudyka, for her partnership during the research trips and chromosome studies, Dr. Vitaly Nechaev for providing living material. The photos made by Yury Semeikin and Elena Volynets are gratefully acknowledged. This study was financially supported by Russian Fund for Basic Research (RFBR), projects 98-04-49455, 01-04-49430, 04-04-49750, 07-04-00610, and 11-04-00240 (to Nina Probatova).

INTRODUCTION: THE MONSOON ZONE IN THE RUSSIAN FAR EAST

The plant cover of the Russian Far East (RFE) is greatly influenced by the position of this region in the monsoon climate zone; the periphery of the largest continent is in contact with the largest ocean, giving rise to a coastline of 22000 km. The monsoon zone covers a considerable part of the RFE, which encompasses the Kamchatka Peninsula, Sakhalin, the Kuril Islands, continental areas and coasts, as well as the islands of the Bering Sea, the Sea of Okhotsk, the Sea of Japan, and the Russian part of the Amur River basin. Generally, at least a quarter of Northern Asia, from the Lena River in the northwest, and Lake Baikal in the southwest, to the Pacific Ocean, is covered by vegetation influenced by the ocean (Box et al. 2001; Safronova and Yurkovskaya 2007).

The monsoons of the temperate zone manifest themselves most clearly in the southern part of the RFE. Many representatives of the RFE monsoon flora are absent in other regions of Russia. The origin of plant cover in the ecotone of the Pacific region of the RFE is determined by the intense influence of East Asian monsoons. The heterogeneity of the monsoon climate in different areas of the RFE contributes to the high variability of the plant cover. The monsoon precipitation increases the contrast in the environment, by producing critical periods that restrict the functional stability of plant communities. The monsoon climate induces the formation of adaptive possibilities. The very special sandbank flora and appropriate plant communities in the Amur River basin exist because of the monsoon climate (Probatova 1961, 1965, 1969a; Woroshilov 1968; Nechaev and Gapeka 1970; Probatova and Sokolovskaya 1981a, Morozov and Belaya 1998a, b; Probatova 2000). The main feature of the Amur River basin is flooding with great fluctuations in water level mainly caused by monsoon rains, which occur twice in the warm season; the highest flood takes place in late summer (August) due to heavy monsoon rains. The Amur River basin is a unique depository of plant diversity, and it represents the northern (northeastern and northwestern) limits of the geographical distribution of many species of southern origin, as well as rare and endangered plant species, that are absent elsewhere in Russia. Some of these species are thermophilic relicts of tropical origin, while some are taxonomic relicts (e.g., *Symphyllocarpus exilis, Bothriospermum tenellum*). Several species of the Amur sandbank flora are endemic to the Amur River basin (e.g., *Beckmannia hirsutiflora, Glyceria leptorhiza, G. amurensis, Agrostis sokolovskajae*). In the monsoon zone the phenomenon of hydrophilous ephemers (e.g., *Dimeria neglecta, Coleanthus subtilis)* has been revealed (Probatova 2000). The influence of monsoons leads to the increased biodiversity of plant communities because of seasonal and annual ecological addition (Ramensky 1938; Akhtyamov 1998).

The Pacific monsoon is the main factor that affects the processes of florogenesis and coenogenesis, and determines the structure and functioning of ecosystems in the RFE (Pshenichnikov 2003; Galanin and Belikovich 2009). In the delimitation of the temperate East Asian phytogeographical region, A.V. Galanin and A.V. Belikovich (2009) drew its limits according to the Pacific monsoon region; the southern limit of this region coincides with the isotherm of the mean January temperature, which ranges from 0 to +5 C. The problems of monsoon climate botany have been discussed in 5 conferences held in Vladivostok, all of which were organized by the Botanical Garden-Institute FEB RAS (Plants in monsoon climate 1998, 2003, 2007, 2009; Monsoon climate plants 2000).

The problem of monsoon zone biodiversity is related to species adaptations, speciation and florogenesis, the formation of plant communities, vegetation dynamics, and population structure. The adaptation of plants to the monsoon climate may lead to significant structural alterations of the vegetative organs (Matyukhin and Manina 2007).

Multiple publications have been devoted to problems of taxonomy, floristics, phytogeography, ecology, plant anatomy and physiology, phytocoenology and nature conservation in the monsoon zone of the RFE, especially the RFE seacoasts and islands, including Probatova (1961, 1965, 1969a, b, 1970, 1971, 1973, 1975, 1976, 1974a, b, 1975a, 1979a, b, c, d, 1981, 1984, 1993, 1997a, b, 2000, 2003b), Seledets (1969, 1970, 1977, 1978a, b, c, 1981, 1988, 2000a, b, c, d, e, f, 2003a, b, 2005 a, b, 2009a, b, c, 2010), Pavlova and Gorovoy (1971), Probatova and Seledets (1980, 1983, 1997), Probatova and Buch (1981), Probatova and Rudyka (1981, 2000), Seledets and

Probatova (1981, 1991, 2001), Probatova and Kharkevich (1983), Tzvelyov (1983, 1985a, b), Borzova et al. (1985), Kozhevnikov (1985, 2001), Kozhevnikov and Korkishko (1985), Barkalov et al. (1986), Barkalov (1987, 1998, 2000, 2009), Roginsky (1988), Rakova (1990), Chubar (1991, 1992, 1994, 1996, 1998, 2008), Probatova and Olonova (1991), Burundukova et al. (1997), Neupokoyeva et al. (1998), Probatova, Seledets et al. (1998), Gorovoy et al. (1999), Pavlova (1999, 2000, 2006), Kozhevnikov and Kozhevnikova (2000a, b, c, 2001a, b, 2007, 2008), Prokopenko (2000, 2001), Barkalov and Eremenko (2003), Chiapella and Probatova (2003), Kozhevnikova (2003), Probatova and Barkalov (2003), Chubar et al. (2004), Burkovskaya et al. (2005), Pshennikova and Berestenko (2006), Barkalov and Yakubov (2007), Smirnov (2009), and Takahashi (2009). The Poaceae have been the taxa of special attention (Probatova 1970, 1973, 1974a, b, 1997, 2003a; Probatova, Barkalov 2003; Tzvelyov 2008, 2009; Tzvelyov and Probatova 2009, 2010a, b). The ecological and phytocoenotic properties of plants in the monsoon zone of the RFE have been the subjects of our studies since the late 1960s.

Many publications contain results of the karyological studies (chromosome numbers) of plant species in the monsoon zone of the RFE: Sokolovskaya and Probatova (1968, 1973a, b, 1974, 1976, 1985, 1986), Probatova and Sokolovskaya (1981a, b, 1982, 1983, 1984, 1988, 1989), Sokolovskaya et al. (1985, 1986), Kozhevnikov et al. (1986), Probatova et al. (1984, 1986, 1989, 1996, 1998, 2000, 2001, 2003, 2004, 2006, 2007, 2009, 2010), Pavlova et al. (1989), Probatova (1989), Probatova and Seledets (1996), Probatova (1998, 2003b, c, 2008), Probatova (2000), Probatova and Rudyka (2003), Probatova and Shatokhina (2007). There is a large collection of voucher herbarium specimens (deposited at the VLA Herbarium, Vladivostok), documenting our chromosome studies. Chromosome numbers are an important source of information on the flora of the RFE, as are taxonomic diversity, karyotaxonomic and evolutionary peculiarities of taxa, karyological polymorphism, taxa of hybrid origin, chromosome numbers and ploidy levels in the species inhabiting various environmental conditions. In Russia, karyological studies of vascular plants have become a matter of considerably heightened scientific interest over the last several decades. The degree of coverage by chromosome counts now makes up over 50% species of the total flora of the RFE.

Chapter 1

ECOLOGY OF THE AREA OF STUDY

The special features of the monsoon zone are clearly manifested on the coasts. Specific characters of coastal plant communities resulted from the specificity of their habitats. The coastal line itself, all ecological factors and a whole complex of factors are extremely changeable.

The main ecological factors in the land-sea contact zone are the strong drying action of winter winds, the absence of snow cover, deep congelation of the ground, lower temperatures during the vegetation period; high humidity during the warm period of the year, high insolation, soil salinization and spraying of plants by sea water, incessant renewal of the substratum and its enrichment by nitrates (birds, sea animals and seaweeds), and poor skeleton soils with a very thin humus layer that does not hold water well. The most vivid processes of the land-sea interaction are observed in rather small shoreland "maritimal belt" (Seledets 1970), that comprises sandy-pebble supralittoral zone, lagoon lakes, estuaries, seaside sands, rocks and slopes of marine terraces influenced by the sea.

THE COASTAL BOTANY

The main peculiarity of the coastal zone under consideration is that all the components of the plant cover, species, populations, plant communities and ecosystems, survive in conditions of permanent environmental changes, high-degree dynamic and intensity of ecological factors. The problem of the land-sea interaction is connected with the most important aspects of botany: species adaptations to coastal environment, speciation and florogenesis, formation of plant communities and vegetation dynamics, population structure.

We proposed considering the complex botanical problems connected with the land-sea interactions as a special part of botany - Coastal Botany (Probatova and Seledets 1998, 1999; Probatova et al. 2003, 2005). The major problems of coastal botany, with special reference to the RFE, are coastal plants and their environment, main ecological factors, diversity of habitats, the "maritime belt", adaptive structures and strategies of plant species, specialization of the coastal halophytes, chromosome numbers and ploidy levels, taxonomy, speciation, new and noteworthy taxa of coastal flora in the RFE, taxa of hybrid origin, phytogeography and the distribution limits of coastal species, plants of intracontinental arid zones along the coasts of the RFE, the anthropogenic component and invasive plant species along the seacoasts of the RFE, ecological differentiation, ecosystems and plant communities, anthropogenic activities on the coasts and the priorities of nature protection in the land-sea contact zone. The consideration of chromosome numbers (different ploidy levels), in connection with species ecology, permits us to assess population stability in the stress environmental conditions on the coasts. The ploidy level could help to define a strategy of surviving and the optimal type of coenotic population.

COASTAL VEGETATION

The vegetation along the seacoasts is physiognomically specific. The specificity of the floristic structure of the coastal vegetation make it resemble desert vegetation. Seacoasts are ecotones of a global rank, so "mixed" plant communities are not rare on the coasts. Species of different ecology can be found within the limits of the same plant community. The rather wide occurence of the xerophytic habitats and, accordingly, xerophilous species, populations and communities is one of the paradoxes of seacoasts.

One of the main features of the coastal plant cover is a variability in the structure and composition of plant communities. A great variety of ecological conditions exists along seacoasts that shape the interrelations between typical coastal species and meadow, forest, even - steppe elements of the flora, which can find suitable habitats there. The coastal plant communities are mixed, chaotic, unstable, unclosed, unordered, and as variable as their composition and structure. They are fragmentary and often occupy restricted areas, forming various combinations in the coastal landscapes.

The low floristic density, low density of plant cover and low degree of relationships between species are typical features of coastal plant communities. The disturbance and near total destruction of coastal plant cover results from storms, landslips and other extreme natural effects and sometimes from intensive human impact.

The coastal part of the monsoon zone in the Southern RFE is characterized by the highest genetic and coenotic variability of vegetation cover (Morozov and Belaya 2009). Biological plasticity, adaptive abilities, and various ecological ranges prove the conformity of species in the monsoon zone to various natural regimes from oceanic coasts to Inner Asia. For this transitional land-ocean zone, wide variability in plant communities is common (Schlothgauer 2009).

On the seacoasts and islands in the Southern RFE (Primorsky Territory) there are original undersized oak forests (*Quercus mongolica*), various prostrate forests, shrub plant communities (*Rosa rugosa, Artemisia gmelinii*) and *Thymus* spp. communities. It is worth noting *Malus mandshurica* as a prostrate life form on coastal rocks, *Taxus cuspidata* on abrupt rocky slopes, *Rosa rugosa* on coastal sands, and various types of plant communities in the spray zone. There are no such plant communities outside of the seacoasts in the RFE. Steppe-like xerophilous vegetation and *Thymus* communities are not rare on rocks or on the pebbled and rubbly grounds of the coastal zone in the RFE.

The interactions of the environmental factors are often very intensive and result in unpredictable consequences, making seacoasts the area of paradoxical ecological situations. Thus, many coastal plants sprayed by seawater or periodically flooded at full sea, have evident features of xeromorphosis, conditioned by withering winds and soils that poorly retain water.

Halophytes occur in the RFE only on seacoasts and rarely on saline grounds because of human impact. Xerophytic habitats, specific plant communities, and coenopopulations are the special features of the sea coastal areas.

In the Southern RFE, sometimes the invasions of dry, hot continental winds take place (Petukhova 2003). Considering these facts, it would be possible to compare the coastal flora with the flora of deserts, but in the coastal habitats of the RFE the C_3-species prevail, while the C_4-species are more common in arid intracontinental regions. This finding could be explained by a more favorable water regime on the RFE seacoasts (Burundukova et al. 1997; Neupokoyeva et al. 1998).

Coastal Flora of the Russian Far East and Its Connections

More than 500 species of vascular plants are found along the seacoasts of the RFE, of which only approximately 60 species can be considered as halophytes (obligate or facultative). The most typical are species of the maritime belt (Seledets 1970). These are typical coastal species that inhabit the shallow water and spray zones, marshes, coastal meadows, coastal rocks, and slopes of marine terraces. The specific characters of the seaside plant cover are the most pronounced in obligate and facultative halophytes, as well as by species that are tolerant of seaside habitats.

The typical seaside plant species in the RFE belong to the following families: *Alliaceae, Apiaceae, Asteraceae, Boraginaceae, Brassicaceae, Campanulaceae, Caryophyllaceae, Chenopodiaceae, Crassulaceae, Convolvulaceae, Cupressaceae, Cyperaceae, Fabaceae, Juncaceae, Juncaginaceae, Lamiaceae, Limoniaceae, Papaveraceae, Plantaginaceae, Poaceae, Primulaceae, Rosaceae, Ruppiaceae, Scrophulariaceae, Zannicheliaceae,* and *Zosteraceae*. The most numerous families of seaside vascular plants in the RFE are *Poaceae* and *Asteraceae*. In the typical coastal flora of the RFE, native species of the following genera occur: *Aizopsis, Allium, Adenophora, Angelica, Arctopoa, Artemisia, Atriplex, Astragalus, Bolboschoenus, Calamagrostis, Calystegia, Carex, Cerastium, Chorisis, Cochlearia, Conioselinum, Deschampsia, Draba, Dracocephalum, Elymus, Festuca, Glaux, Glehnia, Halerpestes, Hedysarum, Hierochloë, Honckenya, Hordeum, Juncus, Kitagawia, Koeleria, Lathyrus, Leymus, Ligusticum, Limonium, Linaria, Mertensia, Orostachys, Papaver, Plantago, Poa, Potentilla, Puccinellia, Ranunculus, Rosa, Rubus, Ruppia, Sagina, Setaria, Stellaria, Scirpus, Scutellaria, Scrophularia, Senecio, Thermopsis, Thymus, Triglochin, Tripleurospermum, Tripolium, Vicia, Zannichellia,* and *Zostera*. The most common are coastal psammophytes and halophytes (e.g., *Honckenya oblongifolia, Mertensia simplicissima, Senecio pseudoarnica, Salsola komarovii, Glehnia littoralis, Puccinellia nipponica, Leymus mollis, Arctopoa eminens,* and *Poa macrocalyx*). Alien plants on the seacoasts of the RFE are not numerous, probably because of limiting factors (salinity): they include the halophytes *Brachyactis angusta, Cakile edentula, Cotula coronopifolia, Hordeum jubatum,* and *Poa subcaerulea*.

The chromosome numbers in coastal plant species of the RFE range from $2n = 12$ (*Linaria japonica, Vicia japonica, Zostera japonica*) to $2n = $ c.100

(*Poa macrocalyx*). Among the species studied, various ploidy levels are found: $2x$, $4x$, $6x$, $8x$ and higher. In the coastal flora of the RFE, diploids ($2x$) prevail. Diploids predominate among halophytes, which testifies to the significant age of this floristic complex. Halophilicity and ploidy levels in species of the coastal flora in the RFE are correlated as follows: diploids (obligate / facultative, %) - 58.8 / 30.7, tetraploids - 32.4 / 46.2, hexaploids and above - 8.8 / 23.1. Chromosome number data obtained for species near seacoasts in the southern part of the RFE suggest that in the contact zone, the ancient parts of distribution areas exist for a number of species, and not only coastal ones. Examples of species with limits of distribution along the RFE seacoasts are: *Artemisia littoricola, A. stelleriana, Chorisis repens, Plantago camtschatica, Puccinellia kurilensis, P. nipponica, Rosa rugosa, Thermopsis lupinoides, Calamagrostis deschampsioides, Glehnia littoralis, Linaria japonica, Scutellaria strigillosa, Dracocephalum charkeviczii, Festuca vorobievii, Koeleria tokiensis, Calystegia soldanella, Erigeron oharae, Kitagawia litoralis, Arctanthemum arcticum,* and *Glaux maritima.* The floristic connections of Northeast Asia and North America are distinct in the coastal plant cover of the RFE. A convincing example of such connections is the grass genus *Arctopoa.* The area of distribution of its most ancient species, *A. eminens,* a coastal halophyte with primitive morphological features, covers a considerable part of the North Pacific coasts; however, its few populations are also represented on the Atlantic Coast of Canada (Labrador). In addition, the species of *Arctopoa* are a "connecting link" between the floras of Siberia, Central Asia and the North Pacific (Probatova 1974b, 1975b, 1995a, 2003a). The close relationships become apparent in various taxonomic groups in the RFE and coastal floras of the Korean Peninsula, China, and Japan. Among coastal species the most typical and abundant are North Pacific (some of them are Northwest Pacific), and the most important position is occupied by the species of Sakhalin-South Kurils-Japan type of distribution (Pan-Japan Sea area).

SPECIAL FEATURES OF THE "LAND - OCEAN" CONTACT ZONE

Halophytes in the flora of the RFE can be found only along seacoasts; these are species that can reproduce under more or less constant saline conditions. Obligate halophytes can have their ecological optimum here.

The specificity of the "continent-ocean" contact zone results from it being the periphery of distribution area for many species of vascular plants (including coastal ones).

In the land-ocean contact zone lie the limits of geographical distribution for many species of RFE vascular plants. For some species, the area of distribution is located along the seaside, and it is one of the characteristic features of this zone. One of the main paths of species migrations is along seashores, namely the spray zone, slopes of marine terraces, marshes and wetlands, and coastal plains (Probatova and Seledets 1999).

The peculiarity of the flora along seacoasts and islands is determined by species connected with the North Pacific area, e.g., *Carex macrocephala, Rosa rugosa, Leymus mollis, Arctopoa eminens,* and *Poa macrocalyx*. Species of Southwestern Pacific distribution (including the Pan-Japan Sea area) are the most typical and numerous among the RFE coastal vascular plants. At the same time, the halophytic coastal floristic complex of the RFE contains some groups connected with floras of arid intracontinental regions. The genera of *Poaceae* that are common in arid regions of Russia *(e.g., Stipa, Leymus, Koeleria, Hordeum,* and *Elytrigia)* are poorly represented (only 1 to 3 species) in the humid RFE climate. Moreover, species of some of these genera are often confined to the seacoasts; examples include *Leymus mollis, L. villosissimus, Koeleria ascoldensis, K. tokiensis, Hordeum roshevitzii, Arundinella hirta,* and *Festuca vorobievii*. In Primorsky Territory (south of the RFE), some coastal species of xerophytic genera such as *Thymus* (e.g., *Thymus ternejicus)* and *Limonium (L. tetragonum)* are present. Some of our data confirm that the species of coastal flora of the RFE, found away from the seacoasts, reflect the ancient status of land-ocean interactions. Limestone paleoreefs characterized by peculiar calcifilous flora occur in the Southern RFE. Isolated relict populations of coastal plants are found in the RFE monsoon zone at a considerable distance from the seashore. *Koeleria tokiensis, Festuca vorobievii* and *Arundinella hirta* are found on the Lozovy (Chandalaz) Ridge (in Eastern Primorsky Territory). *Thermopsis lupinoides* is found on the extreme south of the Kamchatka Peninsula (lakesides of Kurilskoye Lake) and in the southwestern part of Primorsky Territory (Khanka Lake), *Arundinella hirta* and *Dracocephalum charkeviczii* are found along the Razdol'naya (Suifun) River, and *Carex kobomugi* is found on the lakeside of Khanka Lake.

In the flora of the RFE seacoasts and islands, species with various chromosome numbers are represented: from $2n = 10$ (*Picris, Paris, Trillium, Fimbristylis subbispicata,* and *Chelidonium asiaticum*) up to $2n = c.110$ (*Trientalis*). Chromosome numbers of the RFE maritime species vary from 2n

= 12 (*Linaria japonica*, *Vicia japonica*, and *Zostera japonica*) up to 2n = c.100 (*Poa macrocalyx*). Chromosome numbers in general are constant within species. Among coastal plants, there are species with 2x, 4x, 6x, 8x, sometimes - with variable ploidy within species, which manifests the evolutionary processes on the seacoasts.

There are many diploids on the RFE continental coasts and islands, including *Rhodiola integrifolia*, 2n = 22 (North Pacific, mountain tundra); *Primula cuneifolia*, 2n = 22 (North Pacific, mountain meadows, nival belt); *Saussurea neopulchella*, 2n = 26 (Amur - Sakhalin, meadows, forest edges); *Cirsium coryletorum*, 2n = 34 (probably, endemic to Primorsky Territory, forest edges); *Heteropappus saxomarinus* and *Erigeron oharae,* both with 2n = 18 (predominantly Korean, coastal rocks); *Paraixeris denticulata*, 2n = 10 (East and South Asia, rocks); *Thymus ternejicus*, 2n = 24 (endemic to Primorsky Territory, seashores); *Glehnia littoralis*, 2n = 22 (North Pacific, seashores, spray zone); and *Peracarpa circaeoides*, 2n = 30 (Northwest Pacific, forests). The polyploid species include *Viola kamtschadalorum*, 2n = 96 (Northwest Pacific, forests); *Poa macrocalyx*, 2n = 42 - over 100 (North Pacific, coastal meadows); and *Arctopoa eminens*, 2n = 42 (predominantly North Pacific, coastal meadows and marshes).

The book "Karyology of the flora of Sakhalin and the Kuril Islands. Chromosome numbers, taxononomic and phytogeographical comments" (Probatova et al. 2007) is the first of a series devoted to studies on chromosome numbers of the flora in subregions of the RFE. It represents a full database of chromosome number information for 536 vascular plant species studied on Sakhalin, Moneron and (or) on the Kuril Islands, obtained during the karyological studies of the island flora since 1960. The book contains 356 species from Sakhalin (23.4 % of the total amount of vascular plant species of Sakhalin), 257 species from the Kurils (18.4 % of the total vascular flora), and 48 species from Moneron Island studied. The chromosome number data are accompanied by information on the ecology and phytogeography of the species as well as their distribution on the islands. For many species, karyotaxonomic and karyo-geographical comments are given. The chromosome numbers are considered in the context of the literature. For 184 species, chromosome numbers were obtained from plants growing in protected natural areas. Among the leading families of the flora studied, the greatest number of species (98) are representatives of the family *Poaceae*. Among them, the polyploids have high prevalence (72 %), which play an important role in evolution. On the contrary, in the next largest family, *Asteraceae*, the prevalence of diploids is pronounced. Chromosome numbers in the majority of species are constant, and only a few species are characterized by intraspecific polymorphism, testifying about the speciation processes. Up to now the chromosome number data are already known in 558 plant species from Sakhalin, Moneron and the Kurils (Probatova, Barkalov et al. 2009).

THE CONTACT ZONE AS THE AREA OF SURVIVAL

The land-ocean contact zone is a natural testing area for many species of intracontinental origin when they reach seacoasts in the process of their expansion.

Plants occupying substrata of sea origin are constantly in a stressful situation. New habitats are forming continuously on the seacoasts under environmental stress conditions. Speciation is most probable where unoccupied ecological niches occur; these are the zones with the most diverse habitats.

There are a number of types of adaptations to the coastal environment, including halophytic succulence (*Honckenya oblongifolia, Aizopsis maximowiczii*), ribbed leaf blades (*Leymus mollis, Koeleria ascoldensis*), dense pubescence (*Glehnia littoralis, Artemisia stelleriana, Erigeron oharae*); shining upper surface of leaves (*Ligusticum hultenii, Senecio pseudoarnica*), reduction of leaf surface and (or) early leaf senescence and further assimilation by the stem (*Poa vorobievii*), and glaucousness of plants that are covered with wax (*Elymus woroschilowii, Leymus mollis, Mertensia simplicissima, Linaria japonica, Puccinellia nipponica,* and *Papaver sokolovskajae*). Humifuse and pillow-like life forms can be observed (*Honckenya oblongifolia, Thymus spp., Astragalus marinus,* and *Setaria pachystachys*). Many plant species have well developed rhizomes or stolons (*Leymus mollis, Chorisis repens, Arctopoa eminens, Poa macrocalyx,* and *Scutellaria strigillosa*). In many coastal plant species, propagation by seeds is often suppressed, in favor of vegetative propagation.

Halophytes of the coastal zone differ from halophytes of the deserts by having heliomorphic leaf structures with elements of halophilous succulence.

Survival also dictates a rhythm of seasonal development, including an essential shift of the phenophase in 1-2 phenological pentades, as well as a peculiarity of the structure of the seaside populations: here, sharp prevalence of one age group here is the rule, rather than the exception.

The phenomenon of isolation of the monsoon populations of intracontinental species is typical of seacoasts. In recent times, more data have been obtained showing that many genera of vascular plants are represented on the seacoasts by special ecotypes. In a number of genera in the RFE (e.g., *Dracocephalum, Dianthus, Adenophora,* and *Thymus*), a distinct isolation of coastal races of more or less widely distributed intracontinental species is revealed. The fact that the coastal floristic complex exists proves seacoasts to be a zone of natural selection of populations with higher biological potential. A large number of coastal species have long been known from the North Pacific coasts, e.g., *Koeleria ascoldensis, Poa almasovii, P. kamczatensis, P. macrocalyx, P. tatewakiana, Setaria pachystachys, Artemisia littoricola, Arundinella hirta, Plantago camtschatica, P. japonica, Deschampsia macrothyrsa, Puccinellia kurilensis, P. geniculata, P. nipponica, Honckenya oblongifolia, Salsola komarovii, Atriplex subcordata, Astragalus marinus, Mertensia simplicissima, Senecio pseudoarnica, Chorisis repens, Rosa rugosa, Leymus mollis, L. villosissimus, Heteropappus saxomarinus, Calamagrostis deschampsioides, Arctopoa eminens, Carex macrocephala, C. kobomugi, C. pumila, Ligusticum hultenii, Glehnia littoralis,* and *Scrophularia grayana*. The following new coastal species have been described in recent decades: *Poa vorobievii, P. zhirmunskii, P. verae, P. dudkinii, Festuca vorobievii, Hierochloл helenae, Dracocephalum charkeviczii, Thymus ternejicus, Adenophora probatovae,* and *Dianthus stepanovae*. There are also coastal ecotypes of widely distributed intracontinental species, e.g., *Festuca rubra, Potentilla fragarioides* and *Platycodon grandiflorus*.

A specific feature of coastal and island flora of the RFE monsoon zone is a large number of diploid species ($2x$). According to our studies (Probatova et al. 2003, 2007; Probatova 2007), almost all typical representatives studied of the flora of the monsoon zone in Sakhalin, the Kurils, and the islands of Peter the Great Bay in Primorye (the Sea of Japan) are of low ploidy levels ($2x, 4x$), proving their relatively low biological potential and high vulnerability.

The ploidy level does not determine an advantage in relation to any specific ecological factor, but a general strategy of survival and a type of population most preferable for it. Diploids ($2x$) are mainly found in the poor soils of the seacoasts, and they occupy secondary phytocoenotic positions on the rich soils. In ecologically intense but stable conditions, the diploid level is optimal (highly specialized taxa). In cases of instable ecological situation, including anthropogenic influences, the polyploids with their high ecological plasticity have an advantage. Tetraploids ($4x$) consist of environmentally tolerant populations. Hexaploids ($6x$) and species with higher ploidy levels are capable of growing in the most variable conditions.

Speciation is most probable where free ecological niches and favorable conditions for speciation appear, in zones with mosaic structure and high degree of distinctions in habitat ecology. Apparently, favorable conditions for hybridogenesis always exist along the seacoasts because the stressful conditions, such as strong winds, storms, destructions of rocks, changes in the coastline, that create an abundance of secondary habitats.

We consider the seacoasts to be a zone of speciation (Probatova 1995b, Probatova and Seledets 1999). Hybridization has resulted in an increase in the variability of populations. Many coastal species are presumably of hybrid origin, including *Leymus mollis, L. villosissimus, Poa macrocalyx, P. tatewakiana, P. turneri,* and *Arctopoa eminens*. In *Poa*, natural hybrids occur in the coastal areas of the RFE (intersectional hybrids *Poa almasovii, P. kamczatensis*). On the islands of the Peter the Great Bay, a natural hybrid of two coastal species - *Rosa rugosa* and *R. maximowicziana* has been found.

For the majority of continental species, when they reach the seacoasts, the contact zone appears to test their viability under very different conditions. There are some examples of coastal races in the RFE, including *Elymus woroschilowii* (a member of the intracontinental *E. dahuricus* aggr.), *Festuca vorobievii* (*F. ovina* aggr.), *Arundinella hirta*, and *Koeleria ascoldensis* (*K. cristata* aggr.). The existence of the coastal floristic complex confirms that seacoasts represent a zone where features providing high biological potential are being selected.

Evolutionary trends may be revealed when comparing ecology of coastal species. Some are native to seacoasts, while others originated in inland areas and migrated to seacoasts in the geological past. It has been discovered (Seledets 1976b), that *Leymus mollis, Arctopoa eminens* and *Poa macrocalyx,* coexisting at the seacoasts, have the ecological positions that coincide only partially and their ecological optima differ significantly. Therefore, a comparative study of *Arctopoa eminens, Leymus mollis* and *Poa macrocalyx* showed that the first and the second species are of coastal origin, while the third one probably came from inland parts of Megaberingia.

Chapter 2

THE FAMILY POACEAE AS A MODEL GROUP

Special attention was paid to *Poaceae* species represented in the flora of the RFE. The *Poaceae* can be taken as a model family in floristic and florogenetic studies for the regions where it has a leading position in the floras. The study of ecological ranges was conducted with both native (indigenous flora) and alien (invasive) species. Coenopopulations of different position within the geographical area of species and in plant cover were studied.

The *Poaceae* family is rich in representatives in the RFE and is the second largest family in the vascular flora of the RFE, totaling 476 species (excluding the cultivated species) from 92 genera (Probatova 2007; Tzvelyov 2008, 2009; Tzvelyov and Probatova 2010). The 10 largest genera in the RFE are *Poa, Calamagrostis, Festuca, Elymus, Puccinellia, Agrostis, Alopecurus, Glyceria, Hierochloë,* and *Trisetum,* altogether they comprise more than 55% of the total number of *Poaceae* species in RFE. Among subfamilies within *Poaceae*, the subfamily *Pooideae* is the most abundant in the RFE.

The *Poaceae* component of the flora of the RFE is characterized by the significant diversity of its composition caused by the great variety of the ecotopes as a result of the combination of mountain and flat relief, a set of climatic factors, and strong flood influence (Amur River basin). About 50% of species and a number of genera, including *Sasa, Dimeria, Zoysia, Miscanthus, Neomolinia, Stenofestuca, Brylkinia, Moliniopsis, Torreyochloa,* and *Hemarthria,* are known in Russia only from the RFE; 117 species of *Poaceae* have been described from the RFE. The richness of *Poaceae* diversity in the RFE is conditioned by the position of this region on the continental edge.

The islands of the RFE are significantly poorer in *Poaceae*, but on the islands, some taxa occur that are absent on the continent. Thus, *Bambusoideae (Sasa), Brylkinia, Moliniopsis, Stenofestuca,* and *Brachypodium* occur only in

South Kurils and Sakhalin (Probatova and Barkalov 2003). The Kurils, extending for approximately 1200 km from south to north, are the real connective link ("the Kuril Bridge") between Kamchatka and Japan. The history of the Kurils flora is greatly complicated because of tectonic processes, repeated connections of these territories, significant climatic differences between two ends of the chain, partial glaciations (North Kurils), the great influence of the Pacific Ocean (and the Sea of Okhotsk), permanent volcanic activities.

Poaceae is one of the largest families in both Sakhalin and the Kurils: in Sakhalin, it shows the second highest level of taxonomic diversity (after *Asteraceae*), and in the Kurils, it is also second highest in taxonomic diversity (after *Cyperaceae*). Though Sakhalin and the Kurils are insular territories and situated near each other, they are quite different in terms of their diversity of *Poaceae*. The Kurils are richer and more diverse in *Poaceae* than Sakhalin. However, the *Poaceae* flora of Sakhalin seems to be more ancient than that of the Kurils, it demonstrates long history of its formation. Most genera of *Poaceae* are common to both Sakhalin and the Kurils, though five genera (*Arundinella, Muhlenbergia, Neomolinia, Panicum,* and *Stenofestuca*) are represented on the Kurils, and three genera (*Cinna, Macrohystrix,* and *Ptilagrostis*) on Sakhalin. On Sakhalin, a number of "continental" *Poaceae* species occur, though many of them are restricted or rare, such as *Agrostis anadyrensis, A. trinii, A. kudoi, Calamagrostis lapponica, Cinna latifolia,* and *Macrohystrix komarovii*. Some of them are most likely relict, e.g., *Achnatherum confusum, Avenula dahurica, Glyceria triflora, G. spiculosa,* and *Poa sibirica;* all of these species are absent on the Kuriles. "Continental" species are increasingly few (relict) in the Kurils; examples of these include *Festuca altaica* (Paramushir Island), *Achnatherum extremiorientale* (Kunashir and Shikotan Islands). Many other species occur in the Kurils but are absent in Sakhalin, such as *Agrostis alaskana, A. exarata, Alopecurus stejnegeri, Calamagrostis litwinowii, C. sesquiflora, Hierochloë kamtschatica, Hordeum brachyantherum,* and *Poa turneri*. These species are common to the North Kurils and the Kamchatka Peninsula. The other species - *Anthoxanthum nipponicum, Brachypodium kurilense, Calamagrostis aristata, C. hakonensis, C. urelytra, Elymus tsukushiensis, Festuca hondoensis, Glyceria depauperata, G. probatovae, Muhlenbergia curviaristata, Neomolinia japonica, Puccinellia nipponica, Setaria pachystachys,* and *Stenofestuca pauciflora* - are common to the South Kurils and Japan. There are some endemic taxa of *Poaceae* on Sakhalin, including *Deschampsia tzvelevii, Festuca limosa, Poa neosachalinensis, P. dudkinii,* and *P. sugawarae,* but almost no endemic

Poaceae are credibly known in the Kurils. The *Poaceae* of the Kurils are peculiar in terms of the abundance of *Agrostis* spp. (especially, in the North Kurils) and their natural hybrids. Even hybrids between species of different sections are not rare (e.g., *A. alaskana x A. flaccida = A. x paramushirensis*). The Schmidt Peninsula (the extreme north of Sakhalin) is characterized by many natural hybrids (especially in *Poa*), which is very peculiar. Xeromorphic genera, like *Koeleria*, are absent in Sakhalin and especially on the Kurils (but they occur in Japan, e.g., *Koeleria tokiensis*). The xeromorphic meadow-grass *Poa angustifolia* occurs on the islands only as an alien plant. Many species distribution limits have been revealed in *Poaceae* on the Kuriles, suggesting a complicated history of the flora formation.

The halophytic component among *Poaceae* in the RFE is composed of coastal plants, though few are obligate halophytes.

The special features of *Poaceae* in the RFE are influenced by Pacific monsoons, a very special hydrological regime (the Amur River basin) with two floods (in the spring and summer), a predominant mountain relief (75% of the RFE territory), and volcanic activity. Most of the indigenous species are by origin oceanic or sub-oceanic; this is also the case for most of the endemics within *Poaceae* in the RFE.

The humid East Asian and Beringian florogenetic centers were the first to form *Poaceae* flora on the northeastern edge of Asia. Here are some centers of taxonomic diversity (probably centers of speciation) for *Poa, Agrostis, Hierochloë, Glyceria, Calamagrostis,* and *Elymus*, as well as large hybrid zones. Floristic migrations have favored the speciation by means of hybridization.

The indigenous subset of the *Poaceae* in the RFE consists of East Asian elements (over 40%). Endemic species occur in the monsoon zone of the RFE in areas of intensive volcanic activities on Kamchatka, the Kurils, and Sakhalin. Some examples are *Poa shumushuensis, P. kronokensis, P. sugawarae, P. uzonica, Deschampsia tzvelevii, Agrostis pauzhetica, A. kamtschatica,* and *Glyceria voroschilovii*. In the Sikhote-Alin Ridge, especially on its east (coastal) mountainside, the endemic species *Agrostis sichotensis, Calamagrostis latissima, C. tatianae,* and *Poa sichotensis* occur, while in the Lower Amur, *Agrostis sokolovskajae* and *Festuca amurensis* are found. From the western coast of the Sea of Okhotsk, the endemics *Poa almasovii, P. golubii* and *P. koniensis* have been described (Probatova 2004, 2006).

The *Poaceae* species involved in the study of ecological ranges are native (indigenous flora) and naturalized (invasive), with various chromosome numbers and ploidy levels. Chromosome numbers for the majority of species are constant, but *Bromopsis pumpelliana, Calamagrostis amurensis, C. brachytricha, C. langsdorffii, Hierochloë glabra, Koeleria cristata,* some species of *Poa,* etc. are characterized by the intraspecific polymorphism testifying about the speciation processes.

The Latin names of the taxa in this book and chromosome numbers are given according to Vascular plants of the soviet Far East (1985-1996) and Flora of the Russian Far East (2006), as well as to Probatova (2007), Tzvelyov (2008, 2009), Tzvelyov and Probatova (2010).

Chapter 3

MATERIALS AND METHODS

AIM OF THE STUDY

The aim of our study in the monsoon zone of the RFE was to produce the original floristic, karyological, ecological and phytocoenological data that are under consideration here. We evaluated the paths of adaptation to the monsoon zone of the RFE in various taxonomic groups, the special features of vegetation cover dynamics as well as the conditions and results of species cohabitation in the land-sea contact zone. We determined the typical coastal species position in the system of ecological factors to analyze the changes caused by the natural dynamics of plant cover or human impact. The results of our study of the RFE monsoon zone floristic diversity may clarify the processes of florogenesis in Northeastern Asia.

MATERIALS

The materials were the results of a field study of the authors in all regions of the RFE federal district: Republic of Sakha-Yakutia, Kamchatsky Territory, Khabarovsky Territory and Primorsky Territory, Amurskaya Region, Magadanskaya Region, and Sakhalinskaya Region. Over 4000 relevйs served as the basis for the modified ecological scales used for ecological evaluation of plant communities (Seledets 2000; Seledets, Probatova 2003). They cover the monsoon zone, and this is the considerable part of the RFE. We also used descriptions of the habitats, additional environmental and biogeographical data, plant successions, characteristics of natural and economic situations in the regions under study as well as data from the literature on ecological scales (Tsatsenkin et al. 1978; Komarova et al. 2003, etc.). A model group was the Family *Poaceae*.

Comparative studies of plant species in oceanic and intracontinental regions were conducted for different types of vegetation. Our material comprised descriptions of coenopopulations in geographical profiles from Inner Asia to the Pacific coast. For comparative studies of coenopopulations, two geographical profiles, approximately 1000 km each, were traced from north to south, first in the continental region from the Lower Lena River (Kyussur village, North Yakutia) to Yakutsk city, and second in the monsoon zone from Khayilino village (North Koryakia) to Petropavlovsk-Kamchatsky city, south of the Kamchatka Peninsula. The comparison was made to examine whether coenopopulations of the same species differ notably when areas differ considerably. Our aims were to reveal trends in coenopopulation variability in different directions (from the Arctic to Inland Asia and from Inland Asia to the Pacific coast) and to reveal ecological factors responsible for the ecological variability of coenopopulations in connection with their position within the geographical area of the species. We studied the ecological properties of coenopopulations, mainly in *Poaceae* species of the genera *Agrostis*, *Arctagrostis*, *Arctophila*, *Avenula*, *Bromopsis*, *Calamagrostis*, *Danthonia*, *Deschampsia*, *Elymus*, *Elytrigia*, *Glyceria*, *Festuca*, *Leymus*, *Melica*, *Phleum*, *Poa*, *Puccinellia*, *Schizachne*, and *Trisetum*.

THE ECOLOGICAL SCALES METHOD

A comparative ecological study includes various methods, among which are instrumental measurements of environmental characteristics, studies on habitats and plant communities, and phytoindication. Our study is based on the method of ecological scales, proposed by L.G. Ramensky (1910, 1938, 1971), who elaborated his Individualistic concept of vegetation cover organization before H.A. Gleason (1917, 1924). The method of ecological scales is one of the branches of the phytoindication.

The Ramensky ecological scales can be applied for ecological evaluation of lands. The method of ecological scales is a principle part of Ramensky's theoretical heritage. The creation and development of this method is an outstanding page in the history of ecological studies. In 1910, at the XII Congress of Russian naturalists and physicians, Ramensky presented his paper "A comparative method for ecological study of plant communities". In this paper he formulated the main items of his doctrine of the vegetation cover continuity. The Ramensky method of ecological scales grew out of his continuity doctrine (Ramensky 1910, 1924, 1937, 1938, 1971). At first, the Ramensky individualistic concept of vegetation (every plant species has its own ecological features) was rejected in pre-revolutionary Russia. Later, it was accepted not only in Russia but throughout the world, especially for agriculture, forestry and nature management.

The Ramensky doctrine was initially rejected for two reasons: 1) the ideas of Ramensky were new and unusual for Russian scientists at the beginning of 20th century, and 2) acceptance of these ideas might force radical changes in the methods of collecting and processing field data. Independently, the American botanist H. A. Gleason (1917, 1924) presented a similar concept that was accepted with great interest. Later, a widely known work of R. Whittaker (1975) confirmed the Ramensky approach. Presently, this concept is recognized in many countries (Sobolev and Utekhin 1973; McIntosh 1975, 1983; Mirkin 1987; Moravec 1989).

Ramensky vigorously insisted that it was necessary to convert from descriptive methods and subjective estimation to quantitative methods. He predicted the transformation of vegetation science of his time from "semi-art" to exact science. These principles have now received recognition. Ramensky is well known as the author of the "standart ecological scales" or ecological scales. They can be used for determination of biological needs of plants (environmental factors and regimes) as well as for estimation of the natural potential of lands.

The book "Ecological estimation of hayfields and pastures based on vegetation study" (Ramensky et al. 1956) contains data for 1500 species of Eastern European vascular plants. This book is intensively used in basic ecological studies and in practice. It initiated the application of ecological scales in many fields of human activities. Further, the ecological scales were elaborated for many regions of the former USSR, and for some adjacent countries. As to the RFE, the book named "Methodical instructions for ecological estimation of the fodder grounds in tundra and forest zones of Siberia and the Far East" was published (Tsatsenkin et al. 1978). Many years of application of the Ramensky ecological scales, in various lanscape zones, proved them useful for solving numerous practical and basic problems. The Russian Land Use Planning State Institute (Rosgiprozem) recommended the Ramensky ecological scales as the official ecological guide throughout Russia. Further, the ecological scales of this type were made by collaborators and followers of Ramensky (Tsatsenkin 1967, 1970; Ramensky, Tsatsenkin 1968;

Tsatsenkin, Kassach 1970; Sobolev 1971, 1975, 1978; Tsatsenkin et al. 1974, 1978) for the Carpatians, Caucasus, Urals, and Central Asia. This method was modified by V. P. Seledets for the RFE monsoon zone. Then the regional ecological scales were elaborated (Seledets 1975, 1976a, b, 1977, 1978a, b, 1980, 1982, 1985, 2000a, b, c, d).

The Ramensky ecological scales became the subject of special studies (Rodman et al. 1972; Sannikova 1972; Sannikova et al. 1972; Samoilov 1973; Ipatov et al. 1974). These studies proved the Ramensky ecological scales to be more informative in comparison than the other ecological scales of many European authors. The requirements of ecological scales have continually increased (Gabeev et al. 1973; Komarova 1992a, b; Bulokhov 1996; Komarova and Atschepkova 2000; Komarova et al. 2003). The application of ecological scales for describing species habitats is generally accepted (Ipatov and Kirikova 1997; Mirkin and Naumova 1998; Bekmansurov and Zhukova 2000; Mirkin et al. 2000).

The Ramensky method of ecological evaluation of lands is very efficient. Based on his method of ecological scales, Ramensky coordinated the characteristics of plant cover with the parameters of the main environmental factors. The comparison of the environmental evaluation in grades of ecological scales with analogous assessment of the tolerance limits allows for a high degree of formalization of the method without laborious experimental studies. At the same time, the method seems to be all-sufficient; the results obtained by its application may be interpreted without involving concepts from adjacent fields of biology. In particular, the simplicity of the method, as well as the possibility of objective evaluation of the state of ecosystems, implies that this method should be generally recognized (Prilutsky 2007).

The ecological studies include various directions, some of which revealed the reactivity and sensitivity of species to ecological factors (Ipatov and Kirikova 2001), the positions of plant species in vegetation cover (Cheremushkina 2002) and the application of regional ecological scales for classification of vegetation (Komarova et al. 2003).

Our experience applying the Ramensky ecological scales in the Pacific monsoon climate zone proved the necessity of developing ecological scales for particular environmental conditions in this area. Since the early 1970s, we have studied RFE vegetation based on the Ramensky concept of ecological scales. We have investigated some basic problems, such as ecological ordination of plant communities, regularities in natural and anthropogenic successions, and ecological estimation of recreation impact on vegetation, with special attention to suburban areas. We used ecological scales to prove our suggestions for the RFE regional system of protected wildlife areas. We elaborated ecological scales for the most common plant species of the RFE (Seledets 1980, 2000c).

The method of ecological scales assumes that the most important feature of the plant cover is an ecological conditionality. Regionalization is the mainstream of contemporary phytoindication. This method is applicable in many cases when it is necessary to ecologically evaluate an area based on its vegetation. It allows us to evaluate different ecological regimes, such as humidity, soil fertility and salinity, granulometric composition of soil, drainage, variability of humidity, soil renewal, shading, and anthropotolerance (intensity of human impact). Some explanations are needed for the term "soil fertility and salinity", which was proposed by Ramensky. It refers to a spectrum of soils, from the poorest to the rich, very rich and saline. Two environmental factors are traditionally connected: a positive factor (the richness in nutrients) transitions to a negative one (the over-concentration of some "salts"), making the soils unfit for plants. In the RFE, the true salinity of soils is rare; it occurs especially in the coastal spray zone. The fertility of soils implies a sufficient concentration of nutrients, though some cases of artificial salinity are caused by chemical agents.

The environmental conditions in the monsoon zone are greatly variable and are characterized by very different ecological regimes. Plants that occur along the seashores on specific substrates, such as sand, pebble, or underdeveloped soils, often suffer from permanent stress effects. The main ecological factor in the monsoon zone is humidity.

The method of ecological scales, proposed by L.G. Ramensky and developed by the authors in the RFE, permits the representation of ecological ranges of taxa to reveal trends in ecological adaptation and estimate the stage of ecological differentiation of certain taxa. As a further development of the ecological scales method, we proposed the concept of the species ecological range (Seledets and Probatova 2007). This concept can be used to understand the ecological diversity of plant species and adaptive strategies in the monsoon zone of the RFE.

COENOPOPULATION STUDY

We consider the problem of plant cover in the monsoon zone at three levels: the regional level (the RFE monsoon zone), the subregional level (the seaside zone, up to the nearest mountain ridges, including the islands), and the local level (seacoast vegetation of the maritime belt). We obtained detailed data concerning the RFE monsoon zone at the subregional and local levels.

A coenotic population, or coenopopulation (the population of a species within a plant community), is the basic notion for phytogeography, phytocoenology and plant ecology. We studied coenopopulations taking into account the concept of metapopulation (Hanski 1998, 1999; Bassargin and Vorobyova 2003a, b). The ecological evaluation of coenopopulations was performed by applying ecological scales. Besides the predominant species, it also implied involving of ecological scales of their surrounding species within the same plant community (see Appendix 1).

We consider the term "coenopopulation" in the meaning of T. A. Rabotnov (1978, 1995) as a totality of plant individuals of a species in a certain plant community. The coenopopulation studies are usually conducted when they are aimed to the biodiversity conservation (e.g., Baikalova 2005; Kandalova 2005; Utyasheva 2005; Seledets and Probatova 2007), but the field of application of such studies may be much larger: from the evolutionary problems to the practical needs of the phytoindication. The interest in coenopopulation studies increases steadily (Coenopopulations ... 1976, 1977, 1988; Zaugolnova 1976; Silvertown 1982; Dynamics ... 1985; Yurtsev 1987; Chumakova 1988; Zlobin 1989, 1996; Bashtavoy 1990; Hanski 1998, 1999; Smirnova et al. 1993; Bassargin 2002;Bassargin andVorobyova 2003a,b; Seledets and Probatova 2007; Seledets 2009, 2010a, b).

The first stage of the study was the description of vegetation in connection to environmental factors. Further, it was the compilation of ecological images ("ecological portraits") of coenopopulations, taking into account the complex of ecological factors: humidity ("H" scale), soil fertility and salinity ("SFS" scale), anthropotolerance, which in most cases in the RFE meant recreational and pasture load ("RPL" scale), granulometric composition of soil ("G" scale), drainage ("D" scale), renewal of soil ("R" scale), shading ("Sh" scale), and variability of humidity ("VH" scale). Names of types of habitats were given taking into account the environmental categories proposed by D.N. Tsyganov (1974, 1976, 1983), in accordance with the ecological scales of L.G. Ramensky (1971).

In the second stage of the study we measured the amplitudes of ecological factors on geographical profiles to identify the ecological factors most responsible for the variability of the coenopopulations in connection with their position in the geographical area of the species. Based on the results of this study, we described and graphically presented the ecological ranges (see

Appendix 2) and ecological niches of the coenopopulations (Seledets 2006, 2009a, 2010b; Seledets, Probatova 2007).

The third stage of the study consisted of comparisons of the coenopopulations. Seacoasts and islands of the RFE provide the unique opportunity to reveal the limits of ecological tolerance for many species. The method of ecological scales allows the measurement of the interval of ecological factors that corresponds to the phenomenon "sea coastal flora" in grades. According to our data (Seledets 2000a; Seledets, Probatova 2003a, 2007), the ecological limits of plant species on the RFE seacoasts, taken from two main factors, are humidity (H) (62 to 99 grades), and soil fertility and salinity (SFS) (6 to 20 grades).

The base of the regional ecological scales is the assessment of coenopopulations position in the field of ecological factors. We started from the point that ecological scales are an important source of information on the role of species in plant cover (Seledets, Probatova, 2003a, b). The main tendencies of species adaptations to specific environmental conditions in the RFE monsoon zone may be revealed by consideration of ecological features of coenopopulations.

We proposed (Seledets 1985, 2000a; Seledets and Probatova 2007) the successional-arealogical classification of coenopopulations (Table 1), a concept based on the position of coenopopulations in succession within the area of the species distribution. Twelve types of coenopopulations were recognized (centrates, peripherates, isolates and endemates), according to the position of the coenopopulation in a successional series of plant communities (climax, successional and pioneer). This classification of coenopopulations was devised primarily for nature conservation problems. Furthermore, this classification proved to be applicable for the description of the species ecological properties because it reveals the phytogeographical and phytocoenological positions of coenopopulations.

The type of coenopopulation is the indicator of ability of coenopopulation to take certain ecological and coenotic position in vegetation. Taking into account to which of these types a coenopopulation belongs, we may make recommendations for legislative acts and practical work for protection of rare and endangered species.

Certain relationships between the geographical position of coenopopulations and the ecological properties of species have been revealed. Coenopopulations in the central part of the species area of distribution (climax, successional and pioneer centrates) generally have the largest adaptive potential.

Table 1. Classification of Coenopopulations

Position of coenopopulations in the species geographic areal	Position of coenopopulations in the plant cover		
	Climax communities	Serial communities	Pioneer groups
Endemate	Climax endemate	Serial endemate	Pioneer endemate
Isolate	Climax isolate	Serial isolate	Pioneer isolate
Peripherate	Climax peripherate	Serial peripherate	Pioneer peripherate
Centrate	Climax centrate	Serial centrate	Pioneer centrate

Peripheral coenopopulations often have very special features and are, therefore, of great evolutionary value. They often manifest rare adaptations. For example, a small cane, *Scirpus lineolatus*, at its northern limit of distribution (Sikhote-Alinsky biosphere reserve, in the north of Primorsky Territory), is able to exist completely underwater for a long period of time. Peripheral coenopopulations (climax, successional and pioneer peripherates), as well as isolated coenopopulations (climax, successional and pioneer isolates), are usually impoverished genetically, and their adaptive potential is low. Endemic coenopopulations (climax, successional and pioneer endemates) are of great evolutionary interest and are valuable as nature protection. Our classification is applicable to finding the most favorable type of protected wildlife areas for certain coenopopulations.

Chapter 4

THE CONCEPT OF ECOLOGICAL RANGE

THE CONCEPT OF ECOLOGICAL RANGE AS AN ADVANCED STAGE OF DEVELOPMENT OF THE ECOLOGICAL SCALES METHOD

The concept of the ecological range (ecorange, ER) of plant species (Seledets and Probatova 2007) is a further development of the method of ecological scales. It is based on ecological conditionality which is the principal property of vegetation cover. The ER concept considers a species as a system of coenopopulations. The ER is not a simple sum of ERs of coenopopulations put together, but it is a complicated system in which the processes of integration and disintegration determine the integrity of the species as a main taxonomic unit and its internal ecological diversity as the necessary and sufficient condition for survival in diverse and changeable environment. The concept of ER of plant species encorporates the phytocoenological, phytogeographical and evolutionary aspects. The ER, as well as geographical distribution, is a historical phenomenon. Ecological range is the important characteristics, practically, of taxa of all levels. We define the ER as a complex of ecological ranges of coenopopulations within the geographical range of the species. The study of ER is important when we have to estimate the tolerance of species to environmental factors. Plant species are not equally tolerant in different parts of the field of ecological factors (Ipatov and Kirikova 2001). Trends of development of ER structure show the most probable directions of ecological specialization. The ER is a part of ecological space between the axes formed by leading ecological factors. Most often these factors are humidity and soil fertility and salinity, although it is also possible to use granulometric composition of soil, degree of drainage, extent of recreational and pasture load, or other factors. We found that each species has its own ER (see Appendix 2). The study of the features of the ER can help to consider many problems of evolution (adaptations of the species to their habitats, ecological differentiation of the species, their relations), phytogeography (ER in different parts of the geographical range of the species), and nature preservation (characteristics of the ERs of rare and endangered species). The comparative study of the ERs in the intracontinental regions of Asia and on the Pacific coast of Russia is of great importance for revealing of the regularities of form and species formation, structure and functioning of various ecosystems.

PARAMETERS OF THE ECORANGE

The ER is characterized (Figure 1) by its dimensions, configuration and orientation between the axes of ecological factors as well as by its definite position in multidimensional ecological space (MES). The ER is a part of MES, and at the same time it is a part of the ecological niche.

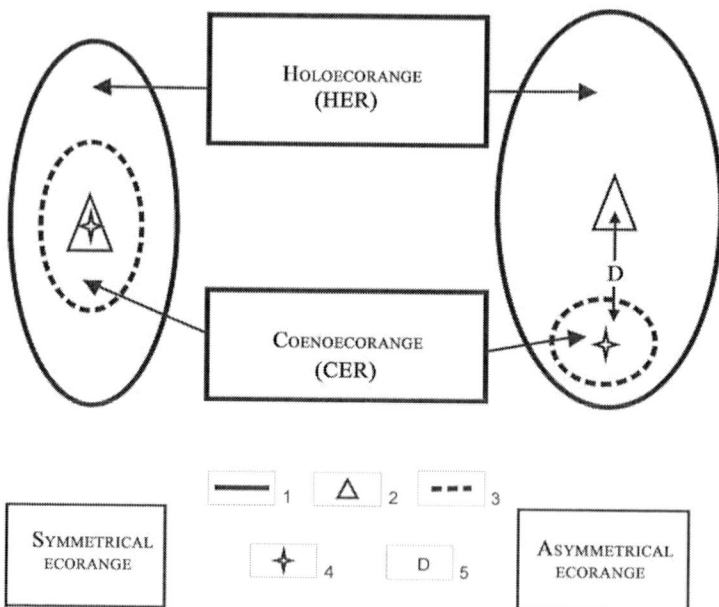

Figure 1. Structure of the Ecorange.

1- limits of the holoecorange
2- center of the holoecorange
3- limits of the coenoecorange
4- ecological optimum (center of the coenoecorange)
5- distance between the center of holoecorange and the ecological optimum (in grades of ecological scales)

Also, we recognize the ER of presence, or the holoecorange (HER), which is the part of ecological space, where the species is able to exist; the ER of dominance, or the coenoecorange (CER), where the species is the most effective as the dominant in the plant community.

The relationship between these two parameters, the CER/HER ratio, indicates the effectiveness of a species in its ER. We also distinguish the ecological optimum and the center of the HER: in some cases, called symmetric ER, these parameters coincide, while in other cases, called asymmetric ER, they do not coincide. The meanings of different features of the ER are provided below.

DIMENSIONS (Figure 2)

Dimensions (size) of ER indicate adaptive possibilities of the species. It may be described graphically and measured in grades of the MES.

The dimensions of the ER define the ecological plasticity of species, i.e., their ability to survive under diverse environmental conditions. When we take into consideration three or more ecological factors, this is the ecological niche. It would be incorrect to consider the ecological niche as a constant property of the species; under certain environmental conditions, the ecological niche may change (Prilutsky 2007).

The dimensions of the ER indicate whether the taxon belongs to an evolutionarily advanced or a regressive taxonomic group. The species in evolutionarily advanced groups are generally characterized by large ERs. *Poa annua* is almost cosmopolitan, and its ER is very large; this species is likely to exist in various environmental situations.

The dimensions of ER result from the multiplication of H x SFS. The maximal ER dimension is theoretically 3600 conditional units: that means 120 (maximal H) x 30 (maximal SFS). In reality, the dimensions of ER are usually less than 1000 units. We recognize three sizes of ER: small (less than 300 units), medium (300-500 units), and large (more than 500 units). ER dimensions, along with other features of the ER, are important for describing ecological niches.

The dimensions of the ER indicate adaptive possibilities of the species. The ER may be shown graphically (see Appendix 2) and measured in grades of environmental space (Table 5).

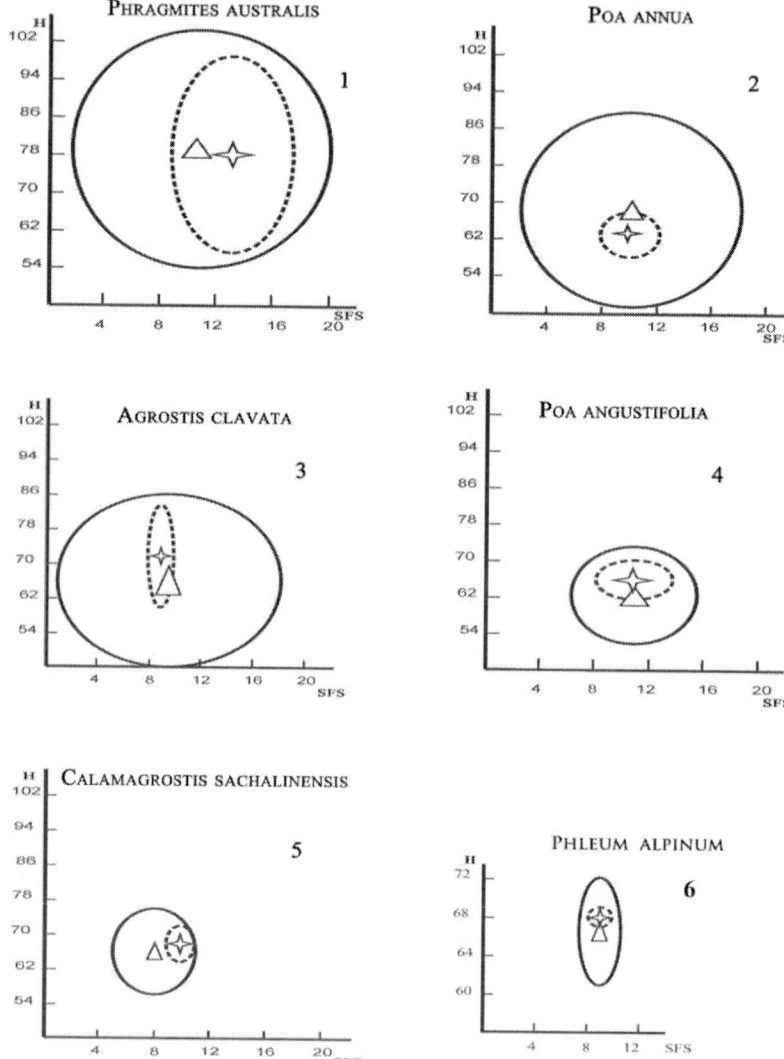

Figure 2. Dimensions of the ecoranges: large - 1, 2, 3; medium - 4, small - 5, 6

CONFIGURATION

The configuration of an ER indicates the trend in ecological adaptation of a taxon. The ER may be stretched along one of the ecological axes, and its shape may vary (round, oval, or elliptical).

POSITION OF THE ECORANGE IN THE FIELD OF ECOLOGICAL FACTORS (Figure 3)

The position of the ER in the MES shows the field of the active ecological adaptation of a taxon. There is considerable diversity of the ER positions in the MES.

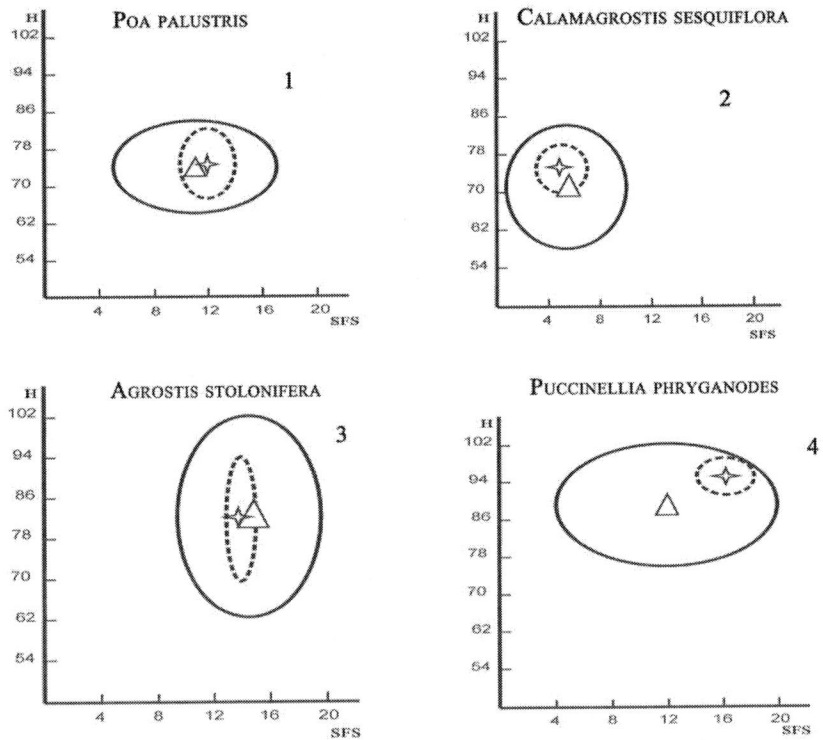

Figure 3. Position of the ecoranges in the field of ecological factors: 1 - central, 2, 3, 4 – peripheral.

ORIENTATIONS OF THE ECORANGES (Figure 4)

The orientation of the ER between the axes of ecological factors shows the area of development of the ER in MES. The ER may be stretched along one of ecological axes. The diversity of ER positions in the field of ecological factors is considerable. The study of the orientation of the ER in MES can reveal trends in the evolutionary processes of various taxonomic groups.

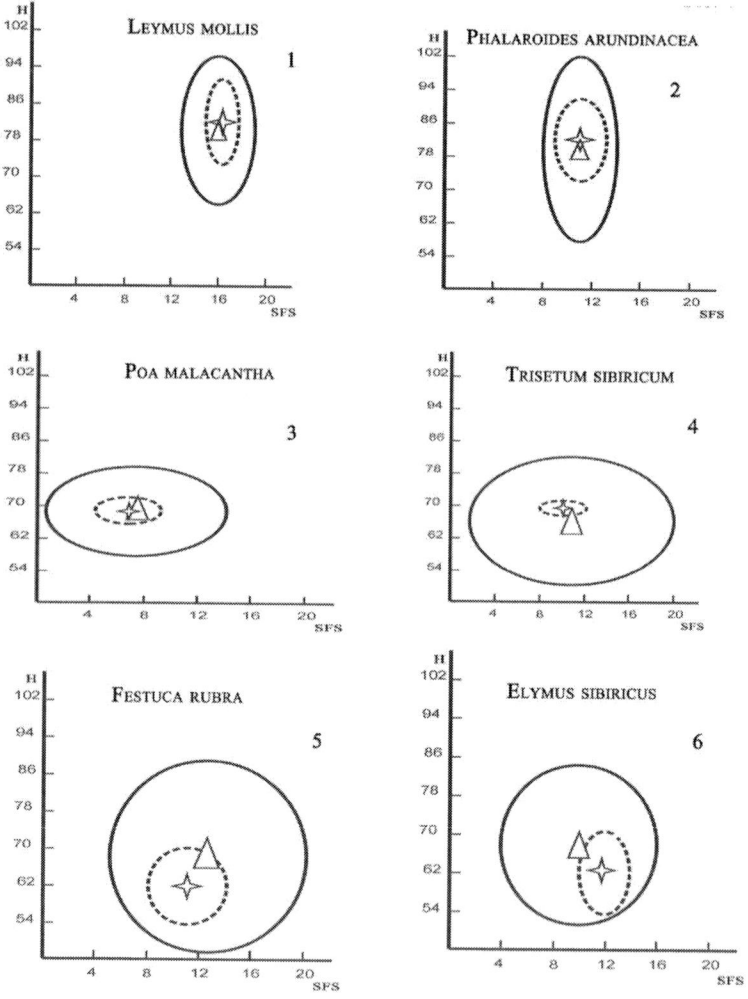

Figure 4. Orientation of the ecoranges: 1, 2 - vertically oriented, 3, 4 - horizontally oriented, 5, 6 - bilaterally oriented.

HOLOECORANGE (Figure 1)

The holoecorange (HER) reveals the ability of taxa to occupy a certain part of the MES.

COENOECORANGE (Figure 1)

The coenoecorange (CER) shows the ecological regime necessary for the species to dominate in plant communities.

ECOLOGICAL OPTIMUM (See on Figures 2- 4)

The ecological optimum is the point in the species ER where the most favorable combination of ecological factors exists. As a result of comparative study of ER and their optima of the RFE coastal species of *Poaceae* two groups of species were revealed. The first group includes mainly the ancient species that are highly adapted to the specific environment of the seacoasts where they realize their biological potential (e.g. *Arctopoa eminens, Leymus mollis,* and *Puccinellia phryganodes*). The second group includes species, that penetrate to the seacoasts most likely owing to competitive relationships (e.g., *Hierochloë glabra* and *Festuca rubra*). This grouping permitted us to approach to the solution of some problems of origin and evolution of the vascular plant ecological groups.

CENTER OF THE HOLOECORANGE (See on Figures 2- 5)

The center of the HER is the most probable (theoretical) ecological optimum if interactions among species are excluded.

DISPOSITION OF THE ECOLOGICAL OPTIMUM AND THE CENTER OF HOLOECORANGE (Figure 5)

The distance between the ecological optimum and the center of the HER shows the direction of the ecological adaptation of a taxon. The coenotic relationships between species lead to the divergence of the ecological optimum and the center of the HER. The more intensive the competition in the plant community, the more this divergence is pronounced.

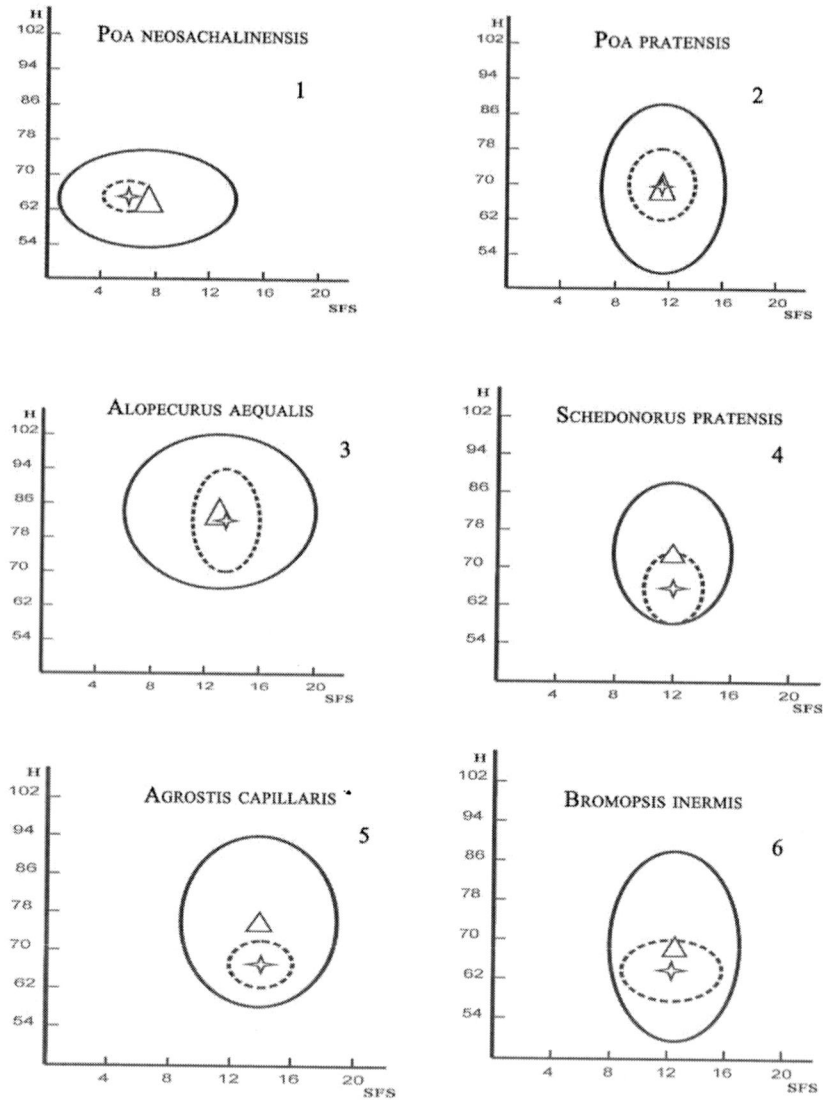

Figure 5. Disposition of the centers of holoecorange and coenoecorange (ecological optimum): close - 1, 2, 3; far - 4, 5, 6.

SYMMETRY OF THE ECOLOGICAL RANGES (Figure 1)

The symmetry of the ER refers to the ecological optimum and center of the HER coinciding. The coincidence of ecological optimum and the center of the HER is typical for successional peripherates (e.g., *Glyceria lithuanica, Avenula schelliana, Arundinella hirta*), and the same is true for successional centrates (e.g., *Poa arctica, P. alpigena, Trisetum spicatum*).

ASYMMETRY OF THE ECOLOGICAL RANGES (Figure 1)

The asymmetry of the ER refers to the non-coincidence or gap between the ecological optimum and the center of the HER. We consider asymmetry of the ER to be the evidence for the primary stage of species adaptation to a new habitat. In some peripherates (*Glyceria spiculosa, G. triflora,* and *Arctopoa subfastigiata*) the ecological optimum is far from the center of the HER. As a rule, it is also far in successional centrates (*Poa macrocalyx, Festuca rubra,* and *Elymus sibiricus*). In the coastal regions mostly asymmetric ERs are typical and indicate a low degree of ecological specialization.

EFFECTIVENESS OF THE SPECIES IN ITS ECOLOGICAL RANGE (Figure 6)

The effectiveness of a species in its ER (CER/HER ratio, %) refers to the degree of ecological adaptation of species. This parameter provides information about the role of a species as a component of plant communities. The highest degree of effectiveness occurs when a species dominates throughout almost the entire ER. A completely effective ER occurs when the HER and the CER coincide, but such species are scarce; far more species are approaching this state (see Appendix 2). The degree of this approach may be measured because when the CER/HER ratio is higher, the ER effectiveness is higher as well.

The position of species in plant communities declares itself through the degree of effectiveness of the ER. Species that dominate in plant communities and ecosystems have ERs of high degree of effectiveness (*Calamagrostis langsdorffii, Arctopoa eminens*). It is possible to recognize grades of effectiveness. We use the following grades of effectiveness of the ER: highly effective ERs, in which the CER occupies more than 50% of the HER, moderately effective ERs, in which the CER occupies from 25 to 50% of the HER, and poorly effective ERs, in which the CER occcupies less than 25% of the HER.

The study of ER effectiveness allows estimation of the viability of coenopopulations, their ability to take the dominant position in plant communities. In dominants (e.g., *Calamagrostis langsdorffii*), the degree of ER effectiveness is considerably higher than in species of secondary positions in vegetation (*Poa malacantha, Trisetum sibiricum*). For *Festuca extremiorientalis, Poa kamczatensis, Glyceria lithuanica,* and *Milium effusum* the very low degree of effectiveness of its ER means that these species are unable to dominate. We suppose that the degree of effectiveness of the ER might be related in some cases to the phylogenetic age of the species: a low degree of effectiveness of ER is characteristic of advanced species. A highly effective ER is a feature of phylogenetically ancient species, e. g., *Arctopoa eminens* which is one of the most ancient representatives of the genus *Arctopoa* that retains many primitive features (Probatova et al. 1984; Probatova 2003).

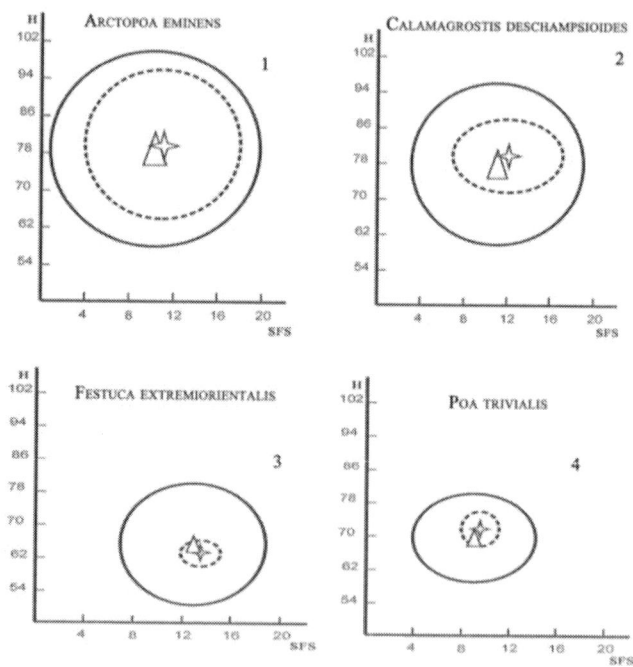

Figure 6. Effectiveness of species in ecoranges: high - 1, 2; low - 3, 4.

The effectiveness of the ER in species with variable ploidy levels increases near the limits of their geographical area. This trend may be illustrated by examples of coenopopulations in serial plant communities: the effectiveness of ER increases from the endemates (*Poa neosachalinensis*, 6x-9x, 2n=42-63) to the centrates (*Poa angustifolia*, 2n=56-64, *P. macrocalyx*, 2n=42-84) and peripherates (*Poa botryoides*, 4x, 6x, 2n=28, 42). Thus, the effectiveness of ER is obviously connected with ploidy levels as well as with position of coenopopulations within geographical area.

ECOLOGICAL PROPERTIES OF SPECIES WITH DIFFERENT PLOIDY LEVELS

The relationships with genetic variability and ecological properties of species became a special field of botanical studies (Grant 1963; Tateoka 1973; Bebbel and Selander 1975; Gray et al. 1979; Ehrendorfer 1980; Stebbins 1984; Rothera and Davy 1986; Bayer et al. 1991; Brochmann and Elven 1992). Polyploidization is considered a major evolutionary process in flowering plants and has attracted much research; thus, ecological studies will contribute substantially to our understanding of the evolution of polyploidy (Bayer 1998). The adaptive significance of polyploidy is that it may allow polyploids to occupy niche space not occupied by diploids. Polyploid complexes are very dynamic, often rapidly evolving, systems.

The meaning of "the ER dimensions" may be illustrated by the RFE *Poaceae* species with different ploidy levels. Connections between the ER dimensions and the position of coenopopulations of species within the area of distribution were demonstrated. The large ERs are found in diploid successional centrates (e.g., *Beckmannia syzigachne, Festuca ovina, Trisetum sibiricum*, 2n = 14).

Tetraploid (4x) species are represented by environmentally tolerant coenopopulations (e.g., *Calamagrostis sesquiflora, Leymus mollis*, 2n = 28). A similar situation was observed for species with higher or variable ploidy levels. The effectiveness of the ER increases from endemates (*Poa neosachalinensis*, 6x-9x, 2n=42-63) to centrates (*P. angustifolia,* 2n=56-64; *P. macrocalyx,* 2n=42-c.100; the same are *Calamagrostis langsdorffii, Poa pratensis, Phragmites australis*). Thus, the effectiveness of the ER is obviously connected to the ploidy levels, especially in species with variable ploidy, and also to the position of coenopopulations within geographical area.

Our studies revealed certain relationships between ploidy levels and ER dimensions (Seledets and Probatova 2005). Among di- and tetraploids 64% have small ERs. Among polyploids of higher levels and species with variable ploidy, 36% have small ERs. Tetraploid *Poaceae* species are of great vitality and tolerance, and the effectiveness of their ERs is the greatest. Examples are the tetraploids *Leymus mollis, Phalaroides arundinacea,* and *Calamagrostis deschampsioides*. Ploidy level is an important indicator of genetic preadaptation to survive in very special environmental conditions.

Typical species of the Southern RFE monsoon flora studied have comparatively low ploidy levels (Seledets and Probatova 1989; Probatova and Rudyka 2000; Probatova et al. 2003, 2007, etc.), e. g., *Artemisia keiskeana, Pterocypsela raddeana, Lamium barbatum, Carpinus cordata, Aster maackii, Dioscorea nipponica, Agastache rugosa, Amphicarpaea japonica* (all 2x), *Sisymbrium luteum* and *Phryma asiatica* (both 4x). Our studies showed that the ecological amplitude of polyploids is generally much broader than that of diploids. However, along the RFE seacoasts, diploids dominate (over 50% of the species studied), especially among halophytes, proving this floristic complex to be ancient (Probatova et al. 1984, 2003c). The study of environmental factors and karyotaxonomic data in the various groups of coastal vascular plants showed that diploids are more common on infertile coastal soils, but they take a secondary position in plant cover on fertile soils because of competitive relationships. Hexaploids (6x) and species with higher ploidy levels are able to exist in a significantly wider amplitude of ecological factors than species of low ploidy levels. Thus, ploidy level is an important indicator of preadaptation of species to ecological stress. In harsh but stable environmental conditions, the diploid level is optimal (specialized taxa). Owing to high ecological plasticity, polyploids take advantage of unstable environmental situations, including human impact. Generally, polyploids in the RFE monsoon zone are characterized by large geographical areas of distribution and ecological amplitude, and a high degree of biological potential, which manifests itself as tolerance to human impact (Probatova et al. 2003c). Correlations of the ERs and ploidy levels in species are especially worth studying.

As an example we consider the ecological tolerance of *Poaceae* species in which we found chromosome races, or variable ploidy levels (*Poa macrocalyx, Calamagrostis langsdorffii, Bromopsis pumpelliana*) in comparison with species of the same genera that were karyologically stable (*Poa tatewakiana, Calamagrostis sesquiflora, Bromopsis canadensis*). The North Pacific coastal species *Poa macrocalyx* is evidently in the process of active differentiation, as confirmed by variable ploidy levels. Asymmetric ER is also a clear sign of favorable prospects to fix in vegetation of the new areas.

Chapter 5

ECOLOGICAL DIFFERENTIATION

TYPES OF ECOLOGICAL DIFFERENTIATION

Different types of ecological differentiation show the diversity of vital strategies, various ways of adaptations to habitats and, certainly, different paths of evolution in plant taxonomic groups. We revealed three types of ecological differentiation of taxa: *linear, diagonal* and *bioriented.*

a) Linear Differentiation

The first type of specialization - *linear* can be demonstrated by the genus *Arctopoa. A. eminens* is the most ancient member of the genus, it is characterized by peculiar halophylous leaf structure and primitive features in spikelet morphology (Probatova 1974, 2003), and it used almost all its adaptive possibilities.The species occurs mainly along the coasts of North Pacific, with few populations on the Atlantic coast of Canada, where its hybrid with *Dupontia psilosantha* has been found (Darbyshire et al. 1992; Cayouette and Darbyshire 1993; Probatova 2003). In *Arctopoa,* the differentiation of the two species studied (*A. eminens, A. subfastigiata*) is clearly manifested; species are located far from one from another along the same axis of humidity. This type of differentiation is linear.

The same situation occurs in *Glyceria*. The ERs of the three species we studied, *G. lithuanica, G. triflora,* and *G. spiculosa,* are located in a line along the same axis of humidity.

Within the genus *Elymus* of the flora of the RFE, three groups of species were revealed according to a combination of ecological factors. According to the classification of Tsyganov (1974), the first group includes oligomesotrophic orthomesophytes (*Elymus macrourus* and *E. kamczadalorum*). The second group includes mesoeutrophic xeromesophytes (*E. mutabilis, E. jacutensis* and *E. sibiricus*), and the third group includes mesoeutrophic orthomesophytes (*E. gmelinii* and *E. kronokensis*). The ecological differentiation is also of the linear type. A hard ecological "packing" has been revealed for *Glyceria* and *Elymus*; these genera are represented in the RFE by a number of species restricted to narrow zones of ecological and phytocoenological spectra.

The ecological differentiation of species is more complicated in the genus *Poa*, the largest genus of *Poaceae* in the RFE: every section of this genus has its own direction of ecological differentiation (also linear). Species of the section *Poa* (s. str.) are mostly mesophytes, while species of section *Malacanthae* occur in orthomesophytic and orthomesotrophic habitats. In *Malacanthae*, which are typical of the RFE, a wide spectrum of ecological adaptations has been revealed.

Representatives of *Malacanthae* are mainly distributed in the North Pacific, they occur in the RFE from high mountains to seacoasts, and their areas of distribution are oriented mostly to the Pacific. The distribution of ecological optima varies considerably in species of the sections *Malacanthae* and *Poa*.

These differences may prove the clear separation of the *Malacanthae* group from the section *Poa* in its former (broader) sense: before, the *Malacanthae* have been included in section *Poa* as a subsection (Probatova 1985) or were not recognized as a separate taxonomic group at all (Tzvelyov 1976). The ERs of section *Stenopoa* are smaller, and the ERs of sections *Macropoa, Nivicolae, Ochlopoa* and *Coenopoa* are the smallest. The comparative age of some closely related species may be determined by the degree of their ecological differentiation.

b) Diagonal Differentiation

In the second type of ecological differentiation, *diagonal*, humidity, soil fertility and salinity are of the same significance (e.g., in *Agrostis* and *Deschampsia*). In *Agrostis* of the RFE, the ecological optima of species in section *Agraulus* (former *Agrostis*) are oriented along the H axis, while in section *Trichodium*, they are oriented diagonally (just between the "H" and "SFS" axes).

c) Bioriented Differentiation

The third type of ecological differentiation is termed *bioriented* and refers to two directions of adaptation to environmental factors. This phenomenon can be illustrated by *Calamagrostis* in the RFE, where the change in direction of ecological differentiation takes place. The first group of *Calamagrostis* species evolved from orthomesophytic and orthomesotrophic habitats to xeromesotrophic and mesoeutrophic ones (*C. sachalinensis, C. brachytricha*). The second group developed from the "point of diversification" to hydromesotrophic and mesoeutrophic habitats (*C. lapponica, C. langsdorffii* and *C. neglecta*). This type of ecological differentiation of species indicates certain trends of evolution in a large taxonomic group. Every line has its own path of environmental adaptation. A wide ecological and phytocoenological spectra also characterize species of *Festuca* in the Northwest Pacific area.

A Global-Scale Ecotone: from the Pacific Coast to Inner Asia

One of the basic problems of phytogeography, ecology and phytocoenology in the land-ocean contact zone is to discover how plants overcome the global scale ecotone, that is, the transition from continental Asia to the Pacific coast. Regularities of the transformation of taxonomic composition of plant communities from the seaside to inland in the monsoon zone were described for the Kamchatka Peninsula since V.L. Komarov (1937); later, many authors studied this phenomenon in detail (Plotnikova and Trulevich 1974; Stepanova 1985; Neshatayeva 1988; Kravchunovskaya et al. 2008, 2009, and others).

The variability of the ERs of the coenopopulations in the land-ocean transitional zone is a new field of study. We compared the Yakutian and Pacific coastal geographical profiles. The ecological optima of many coenopopulations in coastal areas are located in more humid conditions than those of coenopopulations in intracontinental areas. This finding suggests the adaptation of coenopopulations in coastal and intracontinental areas occurred in different ways. While migrating from intracontinental subregions to coastal ones, the species produce various coastal ecotypes.

The largest ERs of the coenopopulations were discovered in Central Yakutia, and the smallest were in the Lower Lena River basin. On the Pacific coast, there are not great differences between the northern and southern coenopopulations. Sometimes, these population sizes are almost equal in the

Northern and Southern RFE. For instance, for some species in North Koryakia (Kamchatka Territory), the ERs of coenopopulations were of almost the same dimensions as in the exrteme south of the Kamchatka Peninsula.

The study of coenopopulations in Yakutia and the Pacific coastal areas of the RFE led us to conclude that the ecological and phytocoenological positions of plant species are determined mainly by the humidity and soil fertility and salinity.

Chapter 6

ECOLOGICAL RANGES OF PLANT SPECIES IN INNER ASIA AND THE PACIFIC REGION

As the territory is too vast, the comparative study of the ERs was conducted in general, by ecological-phytocoenotic groups. We selected within each ecological-phytocoenotic group two subgroups of species: those from Inner Asia and those from the Pacific region.

Each subgroup is illustrated below by few examples. However many species are not typically continental or oceanic, and they may occur in intracontinental areas and in the coastal area as well (e.g., *Festuca rubra, Poa palustris*).

We consider the species with wide geographical distribution to be plants of the monsoon zone, when their ecological and phytocoenological positions become more diverse in direction from inland areas with continental climate to the monsoon zone.

The comparative study takes into account the interspecific relations in plant communities, allowing conclusions separately for each ecological-phytocoenotic group.

MIXED AND BROADLEAF FORESTS

We revealed (Seledets 2006; Seledets and Probatova 2007) that typical representatives of both regions - Inner Asia and the Pacific region, in the mixed and broadleaf forests are characterized commonly by large ERs, which are in the Pacific region mostly asymmetric. This finding indicates a low degree of specialization of species. Examples:

A. Inner Asia

Poa nemoralis L. Holarctic. 2n=6x=42 (hereafter - see in Probatova 2006, 2007; Probatova et al. 2007). Its ER (see Appendix 2) is large, nearly symmetric, and very poorly effective.

B. Pacific Region

Achnatherum extremiorientale (Hara) Keng. East Siberia - Far East. $2n = 2x = 24$. Its ER (see Appendix 2) is very large, asymmetric, and moderately effective.

Festuca extremiorientalis Ohwi. South Siberia - Far East. $2n = 4x = 28$. Its ER (Figure 6) is large, clearly asymmetric, very poorly effective.

SMALL-LEAF FORESTS, FOREST EDGES, AND CLEARINGS

The differences between the intracontinental and Pacific regions in this ecological-phytocoenotic group are expressed as in the previous group, but they are not as significant.

A. Inner Asia

Elymus gmelinii (Ledeb.) Tzvel. Central Asia - South Siberia - Far East. $2n = 4x = 28$. Its ER (see Appendix 2) is medium-sized, asymmetric, and very poorly effective.

Trisetum sibiricum Rupr. East Europe - NW America. $2n = 2x = 14$. Its ER (Figure 4) is very large, clearly asymmetric, and very poorly effective.

Spodiopogon sibiricus Trin. East Siberia - Far East. $2n = 6x = 42$. Its ER (see Appendix 2) is very large, highly asymmetric, and very poorly effective.

B. Pacific Region

Bromopsis canadensis (Michx.) Holub. North Pacific. $2n = 2x = 14$. Its ER (see Appendix 2) is very large, highly asymmetric, and very poorly effective.

Calamagrostis brachytricha Steud. Amur - Japan. $2n = 6x, 7x, 8x = 42, 49, 56$. Its ER (see Appendix 2) is large, asymmetric, and poorly effective.

Elymus kamczadalorum (Nevski) Tzvel. West Pacific. $2n = 4x = 28$. Its ER (see Appendix 2) is medium-sized, asymmetric, and moderately effective.

Poa skvortzovii Probat. Amur - Korea. $2n = 4x, 5x, 6x, 8x = 28, 35, 42, 56$ (Probatova 2007; Probatova et al. 2010). Its ER (see Appendix 2) is large, slightly asymmetric, and poorly effective.

Schizachne komarovii Roshev. Okhotia - Kamchatka (endemic). $2n = 2x = 20$. Its ER (see Appendix 2) is large, symmetric, and moderately effective.

WET AND MOIST MEADOWS

The results of the study show that ERs of the intracontinental species of plants may be very large and vary greatly in size. They are usually elongated (often significantly) along the gradient of humidity, asymmetric, and poorly effective.

Large and very large ERs with low variability in size are typical of the species in oceanic region. They are evenly elongated either along both gradients of H and SFS, or primarily along the gradient of SFS. They are mostly asymmetric, with the degree of asymmetry growing near the coast. The degree of effectiveness of ER varies significantly.

As compared with the intracontinental regions, the ERs of wet and moist meadows at the seacoasts are larger, asymmetric, and oriented along the SFS axis. The species widely distributed in both intracontinental regions and in the zone of influence of Pacific monsoon climate (e.g., *Poa palustris*) comprise a natural transition group.

A. Inner Asia

Phalaroides arundinacea (L.) Rausch. Holarctic. $2n = 4x = 28$. Its ER (figure 4) is large, asymmetric, and moderately effective.

Poa alpigena (Blytt.) Lindm. Holarctic. $2n = 8x - 10x = 56, 60, 70$, etc. Its ER (see Appendix 2) is very large, slightly asymmetric, and moderately effective.

Poa palustris L. Holarctic. $2n = 4x = 28$. Its ER (figure 3) is small, slightly asymmetric, and moderately effective.

B. Pacific Region

Agrostis anadyrensis Socz. East Siberia - Northwest America (Alaska). $2n = 8x = 56$. Its ER (see Appendix 2) is medium-sized, asymmetric, and very poorly effective.

Alopecurus glaucus Less. Siberia - North America. $2n = 14x, 16x = >99, 112, c.120$. Its ER (see Appendix 2) is large, symmetric, and very poorly effective.

Calamagrosis langsdorffii (Link) Trin. East Europe - North America. $2n = 4x - 10x = 28, 42, 56, 70$. Its ER (see Appendix 2) is very large, slightly asymmetric, and highly effective.

PLANT COMMUNITIES OF RIVERBANKS, LAKE SHORES, AND BROOKS

The study shows that the plant species of banks and lakesides on the Pacific coast usually have large, asymmetric and poorly effective ERs.

A. Inner Asia

Agrostis clavata Trin. Eurasia - North America. $2n = 6x = 42$. Its ER (Figure 2) is large, clearly asymmetric, and very poorly effective.

Alopecurus aequalis Sobol. Holarctic. $2n = 2x = 14$. Its ER (Figure 5) is large, symmetric, and moderately effective.

Elymus jacutensis (Drob.) Tzvel. Siberia - North America. $2n = 4x = 28$. Its ER (see Appendix 2) is large, very asymmetric, and very poorly effective.

Setaria viridis (L.) Beauv. Nearly cosmopolite. $2n = 2x = 18$. Its ER (see Appendix 2) is very large, symmetric, and moderately effective.

B. Pacific Region

Agrostis scabra Willd. North Pacific. $2n = 6x = 42$. Its ER (see Appendix 2) is large, asymmetric, and moderately effective.

Glyceria lithuanica (Gorski) Lindm. Eurasia. $2n = 2x = 20$. Its ER (see Appendix 2) is medium-sized, symmetric, and very poorly effective.

Poa shumushuensis Ohwi. North-West Pacific (endemic of the Russian Far East). $2n = 4x = 28$. Its ER (see Appendix 2) is large, asymmetric, and poorly effective.

Zizania latifolia (Griseb.) Stapf. Amur - Japan. $2n = 2x = 30$. Its ER (see Appendix 2) is medium-sized, symmetric, and poorly effective.

BOGGY MEADOWS AND BOGS

In boggy meadows and bogs of the oceanic territories, the ERs are oriented along the H axis, and asymmetry is typical for them. The CER occupies a peripheral part of the HER for *Calamagrostis angustifolia*.

The first thing to note when studying the ERs of this ecological-phytocoenotic group is the differences in size and its variation in intracontinental species. The ER can be very large (*Arctophila fulva*) or very small (*Hierochloë pauciflora*).

The orientation of the ERs of almost all species is clearly expressed along the gradient of humidity. The ERs are nearly always clearly symmetric. The ERs of species on oceanic territories are much larger, in many species they are oriented along the SFS axis and are clearly asymmetric.

A. Inner Asia

Alopecurus brachystachyus Bieb. East Siberia - Amur. $2n = 14x, 16x = c.120$. Its ER is medium-sized, symmetric, and moderately effective.

Arctophila fulva (Trin.) Anderss. Holarctic. $2n = 6x = 42$. Its ER (see Appendix 2) is very large, slightly asymmetric, and moderately effective.

Calamagrostis neglecta (Ehrh.) Gaertn., Mey. et Sherb. Holarctic. $2n = 4x, 10x = 28, 70$. Its ER (see Appendix 2) is large, slightly asymmetric, and effective.

Hierochloë pauciflora R.Br. East Europe - North America. $2n = 4x = 28$. Its ER is very small, symmetric, and moderately effective.

B. Pacific Region

Beckmannia syzigachne (Steud.) Fern. East Europe - North America. $2n = 2x = 14$. Its ER (see Appendix 2) is large, slightly asymmetric, and highly effective.

Calamagrostis angustifolia Kom. Okhotia - Amur. $2n = 4x = 28$. Its ER (see Appendix 2) is large, clearly asymmetric, and moderately effective.

WET ROCKS AND STONY, RUBBLY, SAND-PEBBLE, AND CLAY EXPOSURES

We showed that on the rocks as well as stony, rubbly, sand-pebble, and clay slopes, the ERs are usually medium sized. There is a significant mismatch between the ecological optimum and the center of the HER. Our data suggest that unlike the plant species of zonal vegetation, the plants of rocks and stony, rubbly, sand-pebble, and clay exposures depend little on whether the climate is oceanic or continental. The ERs of these species do not differ much, but the regularities revealed earlier are observed here as well. Ecological features of these species are more clearly revealed if we build the ERs on such ecological axes as granulometric composition of substrate, drainage, and variability of humidity.

A. Inner Asia

Elymus confusus (Roshev.) Tzvel. Siberia - Far East. $2n = 4x = 28$. Its ER is very large, symmetric, and moderately effective.

E. sibiricus L. East Europe - North America. $2n = 4x = 28$. Its ER (figure 4) is medium-sized, highly asymmetric, and poorly effective.

B. Pacific Region

Avenella flexuosa (L.) Drejer. Nearly Holarctic. $2n = 4x = 14$. Its ER (see Appendix 2) is medium-sized, highly asymmetric, and very poorly effective.

Poa neosachalinensis Probat. Sakhalin (endemic). $2n = 6x, 8x, 9x = 42$ 56, 63, 64. Its ER (figure 5) is medium-sized, asymmetric, and poorly effective.

Trisetum molle (Michx.) Trin. East Siberia - North America. $2n = 4x = 28$. Its ER (see Appendix 2) is large, clearly asymmetric, effective.

DRY ROCKS, SCREES, AND OUTCROPS

A. Inner Asia

Achnatherum sibiricum (L.) Keng. ex Tzvel. South Siberia - Amur. $2n = 4x = 24$. Its ER is medium-sized, clearly asymmetric, and moderately effective.

Calamagrostis purpurascens R. Br. East Siberia - North America. $2n = 4x, 6x = 28, 42$. Its ER (see Appendix 2) is small, symmetric, and very poorly effective.

Festuca altaica Trin. Siberia - Far East. $2n = 4x = 28$. Its ER (see Appendix 2) is small, nearly symmetric, and poorly effective.

Poa stepposa (Kryl.) Roshev. East Europe - Siberia - Far East. $2n = 4x, 6x = 28, 42$. Its ER (see Appendix 2) is medium-sized, symmetric, and extremely poorly effective.

B. Pacific Region

Calamagrostis sesquiflora (Trin.) Tzvel. North Pacific. $2n = 4x = 28$. Its ER (figure 3) is medium-sized, clearly asymmetric, and poorly effective.

Melica turczaninowiana Ohwi. East Siberia - South Far East. $2n = 2x = 18$. Its ER (see Appendix 2) is medium-sized, slightly asymmetric, and very poorly effective.

Poa kamczatensis Probat. Okhotia - Kamchatka (endemic). $2n = 7x, 8x = 49, 56$. Its ER (see Appendix 2) is medium-sized, symmetric, and very poorly effective.

P. malacantha Kom. North Pacific. $2n = 6x, 8x, 9x, 10x, 12x = 42, 56, 62, 63, 70, 72, 76, c.80$. Its ER (figure 4) is medium-sized, nearly symmetric, and poorly effective.

DRY AND STEPPE MEADOWS

The ERs of the intracontinental plant species differ in size, from small (*Avenula schelliana*) to large (*Koeleria cristata*). Usually elongated along the gradient of H, they are often symmetric and effective.

The ERs of oceanic territories are usually elongated along the SFS gradient, asymmetric, and effective to a lesser degree than the ERs of Inner Asia.

Fragments of the steppe communities and steppefied meadows are not typical of the oceanic regions and are represented by species for which the basic part of geographic range is situated in continental regions (*Agrostis trinii, Cleistogenes kitagawae, Festuca ovina*).

A prevailing trend, as compared with intracontinental regions, is that the ERs in the Pacific coast are larger, more often asymmetric and poorly effective, in many cases oriented along the SFS axis, rather than the ecological axis H.

A. Inner Asia

Agrostis trinii Turcz. Siberia - Far East. $2n = 4x = 28$. Its ER (see Appendix 2) is large, nearly symmetric, and effective.

Avenula schelliana (Hack.) Holub. East Europe - Asia. $2n = 2x = 14$. Its ER (see Appendix 2) is small, symmetric, and effective.

Cleistogenes kitagawae Honda. South Siberia - Amur. $2n = 4x = 40$. Its ER is large, asymmetric, and poorly effective.

Koeleria cristata (L.) Pers. Holarctic. $2n = 2x, 4x = 14, 28$. Its ER (see Appendix 2) is large, nearly symmetric, and very poorly effective.

B. Pacific Region

Arundinella anomala Steud. East Siberia - Amur - Japan. 2n = 2x, 4x = 34, 34-36, 36. Its ER (see Appendix 2) is small, asymmetric, and poorly effective.

Danthonia riabuschinskii (Kom.) Kom. Okhotia - Kamchatka (endemic). 2n=2x, 4x = 18, 36. Its ER (see Appendix 2) is large, symmetric, and poorly effective.

Miscanthus sacchariflorus (Maxim.) Benth. Amur - Korea. 2n = 4x = 40. Its ER (see Appendix 2) is medium-sized, slightly asymmetric, and poorly effective.

MARITIME BELT

We found that in the maritime belt, the coastal sands and pebbles are dominated by specialized combinations of ecological factors. The plant species of coastal sands and pebbles could not be continental by definition; the basic characteristics of their ERs differ greatly. The ERs are mostly asymmetric, and the degree of their effectiveness generally is moderate.

Arctopoa eminens (J. S. Presl) Probat. North Pacific. 2n = 6x = 42. Its ER (figure 6) is large, nearly symmetric, and highly effective.

Arundinella hirta (Thunb.) Tanaka. Pan-Japan Sea area. 2n = 2x, 4x = 34, 34 -36, 36. Its ER (see Appendix 2) is medium-sized, symmetric, and moderately effective.

Calamagrostis deschampsioides Trin. East Europe - North America. 2n = 4x = 28. Its ER (figure 6) is large, asymmetric, and moderately effective.

Leymus mollis (Trin.) Pilg. North Pacific. 2n = 4x = 28. Its ER (figure 4) is small, asymmetric, and moderately effective.

Poa macrocalyx Trautv. et C. A. Mey. North Pacific. 2n = 6x - 16x = 42, 56, 63, 70, 84, c.100, etc. Its ER (see Appendix 2) is very large, highly asymmetric, and moderately effective.

Puccinellia phryganodes (Trin.) Scribn. et Merr. Holarctic. 2n = 2x, 4x = 14, 28. Its ER (figure 3) is large, extremely asymmetric, and poorly effective.

In general, in Inner Asia the species with large and medium-sized ERs are four times more than with small ones; the symmetry of ERs as well as the effectiveness of ERs is of no great significance as characteristic features. As to the Pacific Region, large and medium-sized ERs are more numerous in 15 times than small ones; asymmetric ERs are three times more numerous than symmetric ones; poorly effective ERs are twice more in number than moderately and high effective ones.

In Inner Asia large and medium-sized ERs are twice less common than in the Pacific region, where small ERs are three times more common. In Inner Asia symmetrical and asymmetrical ERs are almost equal in number; in the Pacific Region asymmetrical ERs are almost three times more in number than symmetrical ones. In Inner Asia poorly effective, moderately and high effective ERs are almost equal in number; in the Pacific Region poorly effective ERS are almost twice more in number than moderately and high effective ones.

Thus, in the whole suite of characteristics the ERs of coenopopulations of the plant species in the oceanic territories differ from the ERs of the plants in intracontinental regions. From continental regions to oceanic territories, the ERs change in dimensions, configuration, and in degree of effectiveness. The ERs in Pacific Russia are usually larger, asymmetric, and much less effective than those in intracontinental regions. In intracontinental regions, ERs with a high degree of effectiveness are common and show high ecological specialization. Symmetric ERs are more common among species of continental Asia; the majority of species studied along the Pacific Coast have more or less asymmetric ERs.

Species of Inner Asia use their ER more efficiently than the majority of species along the Pacific coast of the RFE. The effectiveness of ERs of most species in Inner Asia is noticeably higher than that of the species on the Pacific coast of Russia.

Along with this general regularity, each geographical and ecological-phytocoenotic group features particular regularities of change of size, configuration, and structure of the ERs.

Depending on location of coenopopulation in geographical range of a species the configuration of the ER changes significantly: in its central part it is oriented along the H axis; on the periphery, especially when it coincides with seacosts, the ER is basically directed along the SFS axis. Depending on the phytocoenosis to which a species belongs, the noted regularity can be distinct or, on the contrary, smoothed. Thus, for example, the species in broadleaf forests change the ER orientation less clearly than in other vegetation types.

As far as the specific seacost habitats and their corresponding plant communities (supralittoral, rocks, estuaries, lagoon lakes) are concerned, the ERs show certain specializations.

The size of the ER increases closer to the Pacific Ocean. This trend is observed in different vegetation communities, but no significant enlargement of ER was revealed in the plant communities of rocks, stony, rubbly, sand-pebble, and clay exposures, or in dry meadows.

In monsoon climate asymmetric ERs dominate, and symmetric ones prevail under continental climate conditions. This pattern is especially typical of the species of communities of wet habitats as well as for the species of small-leaf forests, edges, and clearings.

Closer to the Pacific coast, the ecological-phytocoenotic positions of species with variable ploidy tend to strengthen. The ERs of species in supralittoral plant communities of the RFE have much in common with the ERs of rocks, screes, and outcrops, and here the tetraploid species dominate. They are the most stable under the conditions of continuous ecological stress, which is typical of the maritime belt.

Chapter 7

ECOLOGICAL RANGES OF INVASIVE SPECIES

Invasive species (alien, introduced) are an essential and constantly increasing component of floristic diversity (MacArthur and Wilson 1967; Gorchakovsky 1979; Mirkin 1984; Mirkin and Solometch 1989; Burda 1991; Kitayama and Mueller-Dombois 1995; Binggeli 1996). As a rule, disturbed habitats have a comparatively high level of biodiversity, owing to many alien plant species that originated far from the Russian Pacific coast. The "secondary" extension of their geographic distribution into the monsoon zone was observed in the RFE. Similarly, we may also suppose the secondary extension of their ERs.

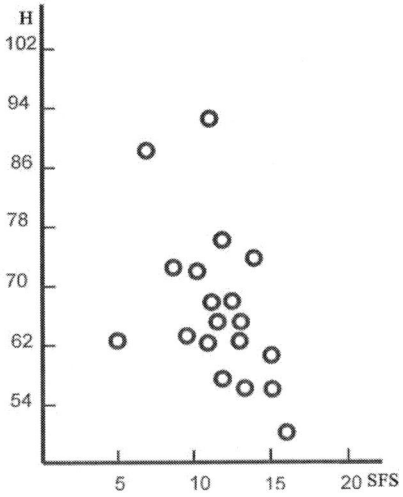

Figure 7. Ecological optima of the *Poaceae* species, that are invasive in the Russian Far East.

Table 2. Description of ecoranges (in grades of ecological scales) of species that are invasive in the Russian Far East

H - humidity, SFS - soil fertility and salinity.
Dimensions of ER are given in conventional units (H x SFS)

Species	Ecological optimum		Extension of ER		Dimensions of ER
	H	SFS	H	SFS	
Agrostis gigantea	81	11	50	15	750
A. capillaris	63	12	32	9	288
Alopecurus arundinaceus	73	13	50	10	500
A. pratensis	72	14	30	13	390
Anthoxanthum odoratum	66	9	25	5	125
Bromopsis inermis	63	13	40	9	360
Dactylis glomerata	63	10	20	8	160
Leymus chinensis	56	12	13	6	78
Phleum pratense	68	12	34	12	408
Poa annua	65	7	61	14	854
P. trivialis	72	10	15	10	150
Schedonorus pratensis	66	11	27	6	162

A significant number of alien species arrive in ports with cargos. Some of invasive species (e.g., *Cotula coronopifolia*, *Hordeum jubatum*, *Brachyactis angusta*, *Atriplex patens*, and *Sonchus asper*) are halophilous and have a succulent leaf structure. Such is the case of the North American sea coastal species *Cakile edentula*, recently found on the seacoasts of southern Primorye and Sakhalin (Chubar 2008; Smirnov 2009). In general, flora of the RFE seacoasts is not rich by the invasive species, owing to a number of limiting factors (for example, salinization) and the intensity of the ecological situation.

The ER may be considered a resource of invasive activity (Table 2, Figure 7). We distinguish three following types of invasion: island invasion, continental invasion, and regional invasion (Seledets 2010a).

ISLAND INVASION

The phenomenon of the island invasion consists of successful reproduction of invasive species on islands, where they occupy new habitats, invade natural plant communities, and expand their distribution area, e.g., *Agrostis capillaris* in Sakhalin (Probatova 1985, 2006, 2007). *Agrostis capillaris* (= *A. tenuis*) is alien to the RFE. However, it is completely naturalized in Sakhalin and, according to properties of its ecological range, it is prospective in the insular territories. The great distance between its ecological optimum and the center of its HER supports this supposition. In the continental parts of the RFE, such species do not occur at all, or they are rare, and there is no reason for their expansion. The most evident feature of the ERs of these species is a low degree of effectiveness.

CONTINENTAL INVASION

There are many cases of invasive species in the continental part of the RFE; nevertheless, most of the species disappear as unexpectedly as they appear. The naturalization of invasive species on the continent is lower than on the islands because the competitive relationships on the continent are tenser. Examples of the continental invasion are *Alopecurus arundinaceus, Arctopoa subfastigiata, Calamagrostis epigeios,* and *Leymus chinensis*. The diversity of the ERs in this group of species is greater than that in the group of the island invasion; differences in the dimensions and effectiveness of the ER are considerable.

REGIONAL INVASION

This situation corresponds to the largest diversity of ecological and phytocoenological situations in the RFE. Species of this group are the most able to invade the plant cover of a new area. Here, we list some examples of this type of invasion in the RFE (all of which are polyploids, 4x, 6x, 8x): *Agrostis gigantea, Alopecurus pratensis, Bromopsis inermis, Dactylis glomerata, Phleum pratense* and *Poa annua*. The reserve of activity in the group of regional invasion is greater than in the island and continental types. The variability of the ERs in this group is more than in other groups. If the distance is considerable, and these species have asymmetric ERs (*Agrostis gigantea, Alopecurus pratensis, Bromopsis inermis* and *Dactylis glomerata*), potentialities of these species are not yet exhausted. *Bromopsis inermis* has great prospects for further invasion into the native flora of the RFE. In some areas, it has taken stable positions in the vegetation. Study of ERs allows for the differentiation of the invasive component of the flora according to the type of adaptation of the species to complex environmental factors. In addition, applying the ER method enables forecasting of the kinds and rates of future invasions of species.

Chapter 8

ECOLOGICAL NICHES

We consider ER to be part of an ecological niche; the niche reflects a species position in multidimensional ecological space (Hutchinson 1965, 1978; Solomon 1969; Parrish and Bazzaz 1976; Harper 1977; Guilarov 1978, 1990; Giller 1984; Shmida and Elener 1984; The population structure ..., 1985; Cody 1991; Malyshev 1993; Palmer and van der Maarel 1995; Rabotnov 1995). The study of ecological niches is one of the most important problems of contemporary phytogeography.

For the descriptions of ecological niches of plant species, we use 8 factors, and the total number of grades for each is indicated in brackets: humidity (H, 120), variability of humidity (VH, 20), drainage (D, 12), soil fertility and salinity (SFS, 30), granulometric soil composition (G, 15), soil renewal (R, 20), shading (Sh, 15), recreational and pasture load (RPL, 10). In comparative studies, the ecological niches seem to be more informative but the ERs have advantages if we know the ecological factors of principal significance. Then we concentrate our attention on them, and the ERs become very useful tool for ecological study.

COASTAL SPECIES ON THE CONTINENT AND ON THE ISLANDS: *LEYMUS MOLLIS* (Table 5, Figure 8)

We compared the coenopopulations of the typical North Pacific coastal species *Leymus mollis* in Sakhalin and in the seashore of continental part of Primorsky Territory (Seledets 2009a). On the continental coast, the ecological niche of coenopopulations of *L. mollis* is broader in all parameters of ecological factors than in Sakhalin. The antropotolerance (RPL) of *Leymus mollis* proves it takes in continental environmental conditions stronger ecological and phytocoenotic positions than on the islands.

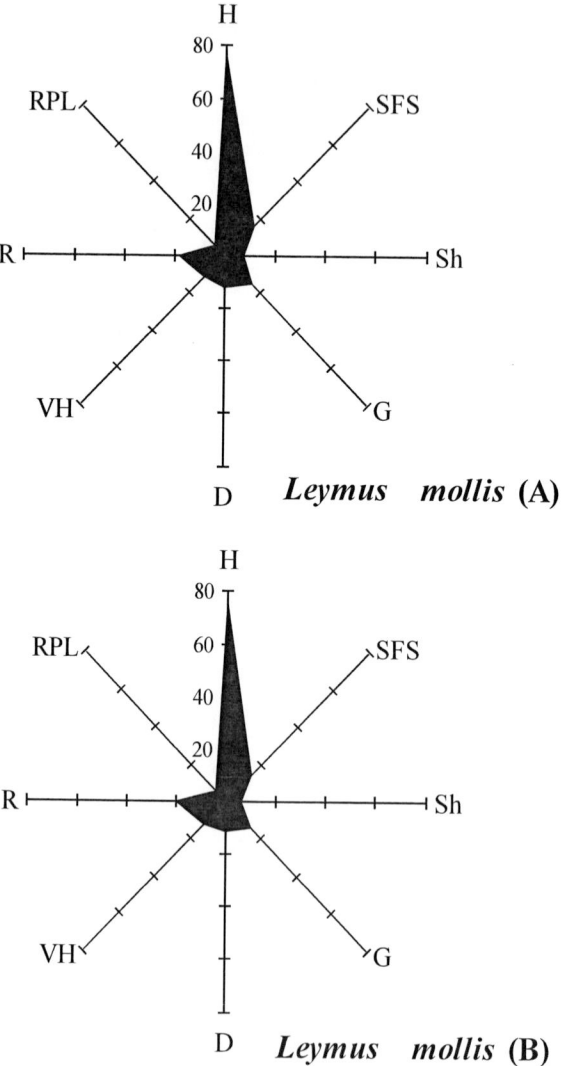

Figure 8. Ecological niches of coenopopulations in different parts of geographical distribution of *Leymus mollis*. A – continental coast of Primorsky Territory; B – Sakhalin.

ECOLOGICAL NICHES ON GEOGRAPHICAL PROFILES:
TRISETUM MOLLE (Table 5, Figure 9)

Trisetum molle is widely distributed in East Siberia, Pacific Russia, Japan, and North America. It occurs mostly on stony slopes, in mountain tundra and forests, and on riverbanks.

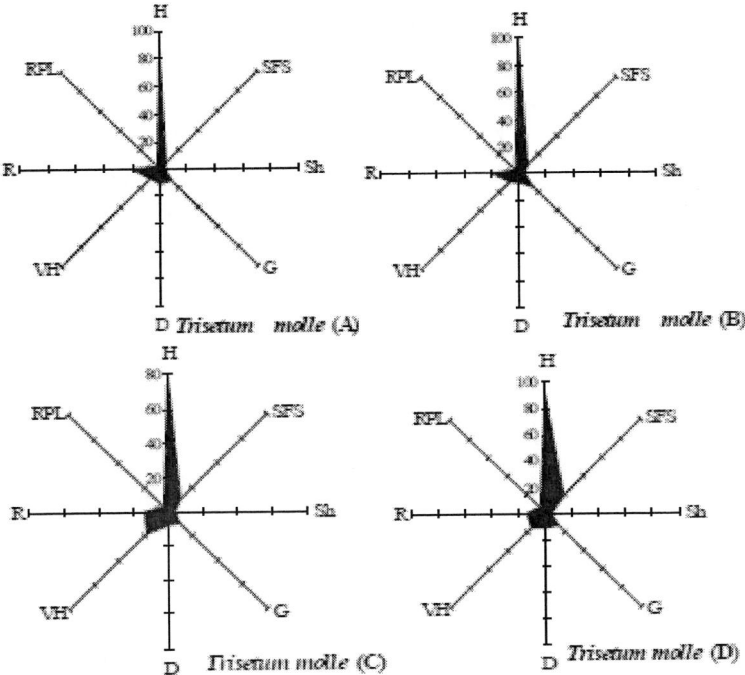

Figure 9. Ecological niches of coenopopulations in different parts of geographical distribution of *Trisetum molle*. A – the north of Yakutia; B – continental part of Kamchatka Territory; C – central part of the Kamchatka Peninsula; D – the south of the Kamchatka Peninsula.

Coenopopulations of *T. molle* from the Lower Lena River basin (Yakutia) and from the northern part of Kamchatka Territory (North Koryakia) as well as from the central and southern parts of the Kamchatka Peninsula were compared.

On the geographical profile from the north to the south of Kamchatka Territory, the ecological variability of coenopopulations increases, especially on the Kamchatka Peninsula, where the ecological amplitude of humidity (H) in the central part is 60-80 grades, but in the south, it is 60-90 grades. The other ecological factors show similar trends: soil fertility and salinity (SFS) in Central Kamchatka is 8-12 grades, while in the south, it is 6-18 grades. Recreational and pasture load (RPL) in the central part is 3-4 grades, while in the south, it is 1-6 grades. Variability of humidity (VH) in the central part of Kamchatka is 5-8 grades, while in the south, it is 2-18 grades. Renewal of soil (R) increased from 10-11 to 1-14 grades. Granulometric composition of soil (G) varied from 5-6 grades in the central part of Kamchatka to 1-15 grades in the south. Drainage (D) varied considerably, from 5-6 grades in the central part of the Kamchatka Peninsula to 1-11 grades in the southern part.

The ecological variability of coenopopulations from the west to the east becomes clear along the geographical profile from East Siberia (Lower Lena) to the northern part of the Kamchatka Territory (North Koryakia): H varied from 65-85 to 63-95 grades, SFS varied from 3-9 to 1-9 grades, and VH varied from 6-12 to 8-16 grades. RPL, R, G and D did not show considerable variation.

Summarizing these data, we can see that in environmental conditions of the Russian Pacific monsoon climate, the ecological variability of *Trisetum molle* coenopopulations is evidently larger than in the continental climate of East Siberia. The comparative study of ecological variability of coenopopulations along the long-distance geographical profiles from north to south (1000 km) and from west to east (2000 km) shows a trend of increases of the ecological amplitudes of all factors. This trend becomes clearer in the diagonal geographical profile, from the northern part of East Siberia to the southern part of the Kamchatka Peninsula.

Comparative study of coenopopulations in Yakutia and in the Pacific coastal areas of the RFE led to the conclusion that ecological positions of plant species are determined mainly by humidity and soil fertility and salinity. In continental subregions, the size of ER of coenopopulations reflects a degree of continentality. The largest ERs of coenopopulations in Central Yakutia (near Yakutsk) has been revealed, the smallest ones - in the Low Lena River basin. In the Pacific coast areas the differences between coenopopulations are not so great, sometimes they are almost equal in the north and in the south: the ERs of grasses in the North Koryakia coenopopulations are of almost the same size as in the extreme south of the Kamchatka Peninsula. It was revealed that ecological optima of many coenopopulations of coastal areas are located in more humid conditions than those of coenopopulations in continental areas. There may be different types of adaptation of coenopopulations in coastal areas versus continental areas: continental coenopopulations are oriented mainly along the H axis, while coastal ones are oriented along the SFS axis.

Ecological Niches 59

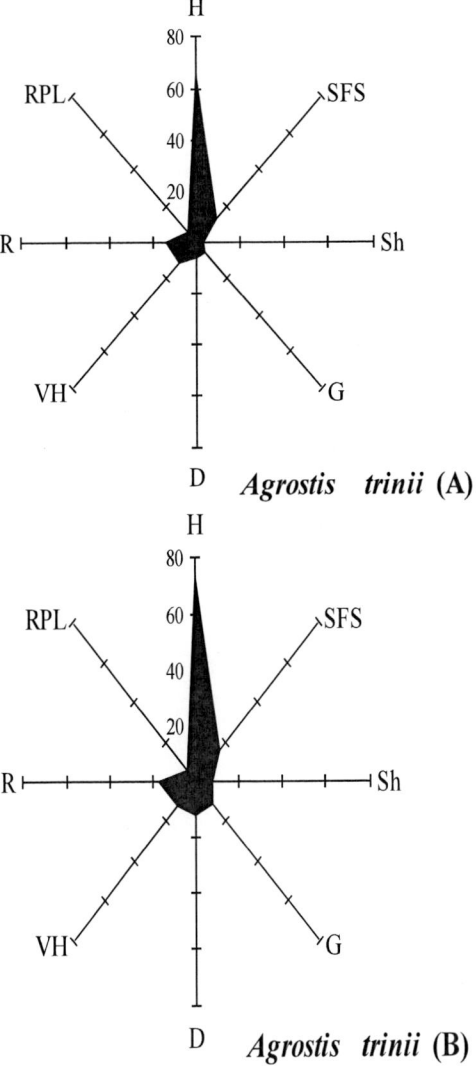

Figure 10. Ecological niches of coenopopulations in different parts of geographical distribution of *Agrostis trinii*. A – continental part of Primorsky Territory; B - coast of the Sea of Japan.

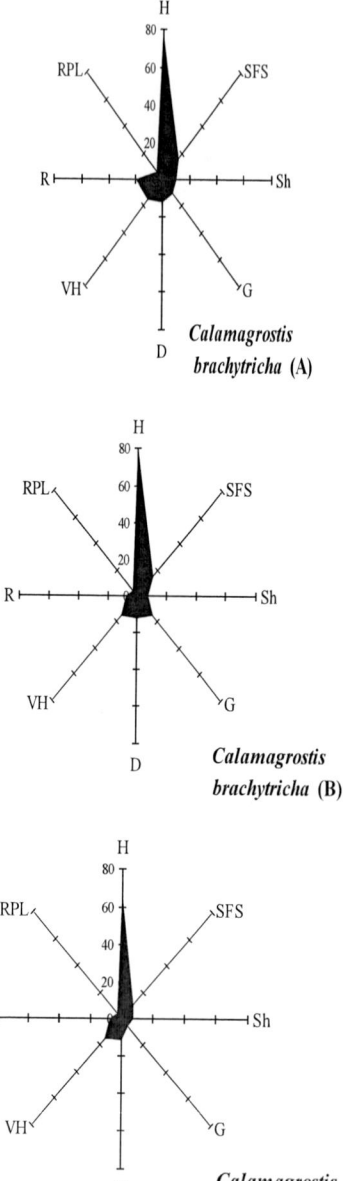

Figure 11. Ecological niches of coenopopulations in different parts of geographical distribution of *Calamagrostis brachytricha* (Primorsky Territory). A – foothills of the East Manchurian Mts.; B – continental coast of the Sea of Japan; C – the islands of Peter the Great Bay.

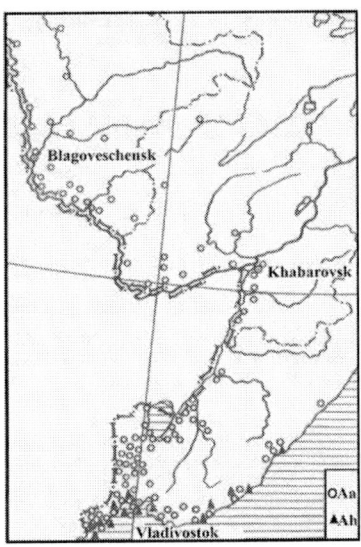

Figure 12. Distribution map of *Arundinella anomala* and *A. hirta* in the Russian Far East.

ECOLOGICAL NICHES OF SPECIES AT THE LIMITS OF THEIR GEOGRAPHICAL DISTRIBUTION: *AGROSTIS TRINII* AND *CALAMAGROSTIS BRACHYTRICHA* (Table 5, Figures 10, 11)

The revealing of the ecological niche at the limits of geographical distribution of species is a special problem of biogeography. New approaches are needed, among them such as the concept of ecological ranges and ecological niches. Taking into account the main ecological factors we studied the transformation of ecological niches in *Agrostis trinii* ($2n = 28$), a mainly Siberian lowland species of dry meadows, and *Calamagrostis brachytricha* ($2n = 42, 49, 56$), a species typical of the East Asian monsoon zone, where it occurs on the slopes in forest edges, and post-forest shrub communities. They are representatives of two closely relative genera, and both reach the Pacific coast in Primorsky Territory. They are mostly continental species as to their geographical distribution. Their ERs were studied in the inland areas, on the islands (Peter the Great Bay) and on the continental coasts of South Primorye, where both of the species are at the eastern limits of their areas of distribution. We showed that the most variable ERs are those of coenopopulations on the Pacific coast. However, for *Agrostis trinii,* the greatest variation occurs in the territory from coastal plains to the bases of mountains, and for *Calamagrostis brachytricha,* the greatest variation occurs from mountain slopes to watershed ridges. We concluded that the ecological niche of *Agrostis trinii* on the Pacific coast of the RFE is much broader than in the continental areas, in all ecological parameters. The broadest ecological niche of *C. brachytricha* is in low mountain belts of Sikhote-Alin and the East Manchurian Mts.; its niche is considerably narrow along the seacoast and minimal on the islands. Continentalization of climate causes continentalization of the ecological niches of these species. Thus, the species becomes increasingly vulnerable as it moves away from its typical continental ecotopes.

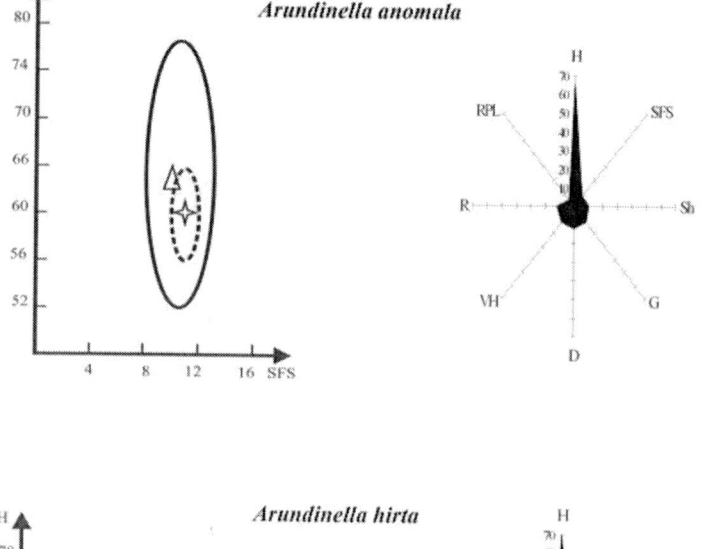

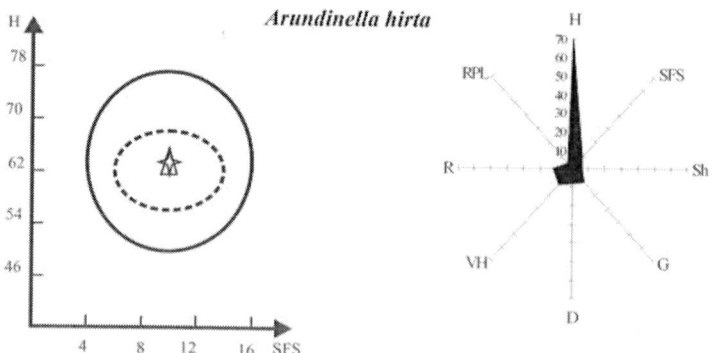

Figure 13. Ecological ranges and ecological niches of *Arundinella anomala* (upper row) and *Arundinella hirta* (lower row), in Primorsky Territory. Ecological ranges are given according to the factors H and SFS.

ECOLOGICAL NICHE AS A DISCRIMINATIVE CHARACTERISTIC: *ARUNDINELLA ANOMALA* AND *A. HIRTA*
(Table 5, Figure 13)

Ecology is of great importance in taxonomic studies. The method of analyzis of ER allows the study of ecological differentiation of taxa as supplementary argumentation for taking taxonomic decisions. As a rule, the ecological and phytocoenological properties of species are taken into consideration when the specific rank of a taxon is under debate, and perhaps also during all following stages of its taxonomic history.

In the flora of the RFE, the genus *Arundinella* (*Poaceae*: *Arundinelleae*) is represented by two closely related species, *A. anomala* and *A. hirta*; sometimes these species are considered subspecies of *A. hirta* s. l. Both are at the northeastern limits of their areas of distribution in the RFE, as well as at the limits of the geographical area of the whole genus. *A. anomala* occurs in Russia from Southeastern Siberia to the Pacific coast, while *A. hirta* occurs along the Pacific coast of Russia. Both are also represented in adjacent countries. The genus *Arundinella* includes approximately 50 species distributed generally in tropical and subtropical regions of both hemispheres, especially in Southeastern Asia. *A. anomala* is common in dry meadows, on rocky slopes, sands and pebbles, up to the lower mountain belt. *A. hirta* s. str. is mainly a plant of the seashores, and in Russia it is represented only in the south of the RFE; it plays an appreciable role in vegetation on the slopes of marine terraces, coastal meadows and pebbles (Tzvelyov 1976; Probatova 1985, 2003a). Both *Arundinella* species are of low value as forage grasses; they are suitable as fodder only before flowering (Tzvelyov 1976).

In the case of *A. anomala* and *A. hirta*, it is important to take into account the following circumstances: the morphological differences between the two species are weak and not always clearly manifested, even some transitions occur. The chromosome numbers of both species are variable: $2n = 34, 34-36, 36$ (see Probatova 2007). These species are not always clearly separated ecologically, e.g., on the seacoasts they may occur on similar habitats. Generally, in the intracontinental areas, *A. anomala* occurs, but on the seacoasts, mostly *A. hirta* is found. However, the geographical distribution of these species in the RFE, which was shown by Probatova (1985, 2003a) in "Vascular plants of the Soviet Far East", differs appreciably (Figure 12). For this reason, it was important to study the ecological and phytocoenological properties by applying the methods of ecological scales and ERs.

Arundinella species have attracted the interest of taxonomists, phytogeographers, and ecologists because they are typical for the monsoon climate zone and can be the guiding points for determining the place of the herbaceous plant communities of the Southern RFE in the global schemes of plant communities and ecosystems (Kurentsova 1952, 1955). P.D. Yaroshenko (1955, 1958) underlined that *Arundilella* species take an essential position in the landscapes of East Asia. Here, he meant zonal formations of the forest-steppe landscape in Northeast China which is characterized by dry stepped meadows, meadow-steppes and typical steppes, oak forests, and *Arundinella* "praries". In the Tsitsikar steppe landscape (Northeast China) the stepped-saline meadows, where *A. anomala* dominates, take a special position among intrazonal formations. In the RFE, there is a zonal formation of the Khanka Plain forest-steppe, that also has oak forests and *Arundinella* "praries". The structure of these plant communities is not steppe, but meadow, because *Arundinella* does not form typical dense tufts, it has short rhizomes. The *Arundinella* meadows occur in Northeast China in a complex of saline meadows which are absent in the RFE aside from the seacoasts.

Accordind to G. F. Patrievskaya (1959a, b), flat piedmonts of the Khanka Plain, higher geomorphological levels of the plain and highlands are occupied by *Arundinella anomala* plant communities for which a complex structure of associations, multi-colored aspect (especially in midsummer), and high species diversity is typical. For these habitats, two maxima of the rapid growth of herbage are characteristic, and two maxima of increasing growth of herbage are typical for vegetation in the monsoon zone.

Considering the meadow vegetation of the Southern RFE from the viewpoint of the monsoon climate, M. Kh. Akhtyamov (1998) examined the general features and ecological differences of *Arundinella anomala* and *A. hirta*. They are similar in that both species are grasses with C_4-metabolism. The author referred them to associations of the union *Arundinellion anomalae* Acht. et al. 1985, sub-union *Miscanthenion sacchariflori* Acht. 1995. These species are representatives of subtropical and tropical floras of East and South-East Asia; they are remnants in the RFE only in those areas where in the period of the summer monsoon maximum (July-August), a humid tropical climate forms, with mean-day temperatures above +22.5 °C and mean-day relative air moisture above 80%. Plant communities of the union *Arundinellion anomalae* are distributed in all regions with monsoon influence. The author considers their variability to increase from the Paleo-Pacific to the Neo-Pacific, where specific features of the *Arundinelletum hirtae* communities are manifested. In the Paleo-Pacific, relict plant communities exist (Akhtyamov 1998).

The aim of our study was to compare the ERs and ecological niches of *A. anomala* and *A. hirta* on the seacoasts, and in the intracontinental areas of the RFE. The area of study is in the southern part of Primorsky Territory, where both of these species occur.

Table 3. Ecological characteristics of coenopopulations and floristic composition of plant communities with *Arundinella anomala* in Primorsky Territory (in grades of ecological scales)

Numbers of relevés groups	1	2	3	4	5	6	7	8	9	10
Humidity	61	61	61	59	58	61	60	61	59	63
Soil fertility and salinity	11	11	11	12	12	11	11	11	11	10
Granulometric composition of soils	8	2	8	4	4	3	13	10	12	4
Drainage	10	11	6	7	10	10	12	12	12	12
Recreational and pasture load	3	3	4	5	4	3	3	3	3	3
Variability of humidity	12	13	7	8	12	12	12	10	11	9
Renewal of the soil	5	6	8	8	4	5	7	4	10	11
Shading	1	2	7	7	9	11	1	1	1	1
Projective cover, %										
Achnatherum extremiorientale	+	-	-	-	-	-	-	-	-	-
Aconitum albo-violaceum	+	-	-	-	-	-	-	-	-	-
Adenophora pereskiifolia	-	-	-	-	-	-	-	-	-	+
Aizopsis aizoon	-	-	-	-	-	-	-	-	-	+
A. selskianum	-	-	-	+	-	-	-	-	+	-
Ajania pallasiana	+	+	-	-	-	-	-	-	-	-
Allium senescens	+	+	-	-	-	-	-	+	-	-
Artemisia gmelinii	5	5	-	5	40	50	20	-	-	+
A. mandshurica	10	-	-	-	-	20	-	-	-	-
A. selengensis	-	-	5	-	-	-	-	-	-	-
A. stolonifera	-	-	-	+	-	-	-	-	-	+
Arundinella anomala	40	10	30	20	10	10	1	5	+	30
Calamagrostis brachytricha	-	-	20	-	-	-	-	-	-	1
Carex nanella	1	10	5	-	20	-	-	-	-	-
C. ussuriensis	-	-	-	10	-	-	-	-	-	-
Cleistogenes kitagawae	10	3	10	-	-	-	3	1	30	-
Clematis hexapetala	-	-	-	-	3	3	1	1	+	-
C. manshurica	+	-	5	-	-	-	-	-	-	-
Dasiphora mandshurica	+	5	-	-	-	-	-	-	-	-

Table 3. (Continued)

Numbers of relevés groups	1	2	3	4	5	6	7	8	9	10
Dioscorea nipponica	-	-	-	-	+	+	-	-	-	-
Dontostemon dentatus	-	-	-	-	+	+	-	-	-	-
Elymus gmelinii	-	-	5	-	-	-	-	-	-	-
E. pendulinus	-	-	+	-	5	20	-	-	-	-
E. sibiricus	-	-	-	-	-	-	-	5	-	-
Eriochloa villosa	-	-	+	-	-	-	-	-	-	-
Festuca ovina	-	-	-	-	-	-	-	1	-	30
Galium boreale	1	+	-	-	+	+	-	-	-	-
Geum aleppicum	-	-	-	-	-	3	-	-	-	-
Gypsophila pacifica	5	3	-	1	+	-	-	1	+	-
Heteropappus hispidus	-	-	-	-	-	-	-	3	5	-
Hypericum attenuatum	-	-	-	-	-	+	-	-	-	-
Iris uniflora	-	+	-	-	-	-	-	-	-	-
Juniperus davuricus	+	-	-	-	-	-	-	-	-	-
J. rigida	+	-	-	-	-	-	-	-	-	-
Kitagawia terebinthacea	+	+	-	-	-	-	-	-	-	-
Koeleria cristata	-	-	+	-	10	-	3	5	-	-
Lathyrus quinquenervius	-	-	-	+	-	-	-	-	-	-
Leontopodium leontopodioides	-	-	+	-	-	+	-	-	-	-
Lespedeza bicolor	5	+	-	-	-	-	-	5	-	3
Lilium buschianum	-	-	-	-	-	+	-	-	-	-
L. distichum	+	-	+	-	-	-	-	-	-	-
Orostachys malacophylla	-	-	-	-	-	-	-	-	+	+
Paeonia obovata	+	+	-	-	-	-	-	-	-	-
Patrinia rupestris	3	+	-	-	-	-	3	1	-	-
Physocarpus amurensis	+	-	-	-	-	-	-	-	-	-
Securinega suffruticosa	1	+	-	-	-	-	3	-	-	-
Selaginella tamariscina	-	-	-	-	-	-	1	+	-	+
Serratula mandshurica	10	-	-	-	-	-	-	-	-	-
Setaria pumila	-	-	+	-	-	-	-	-	-	-
Silene foliosa	-	-	-	-	-	-	1	-	-	-
Sophora flavescens	-	-	-	-	-	-	5	-	-	-
Spodiopogon sibiricus	10	-	5	-	-	-	1	-	+	-
Stipa baicalensis	-	-	5	-	3	-	+	20	+	-
Syneilesis aconitifolia	-	-	-	+	+	+	-	-	-	+
Thymus disjunctus	-	-	-	-	-	-	50	30	30	-
Vicia amurensis	-	-	-	-	-	3	-	-	-	-

Field investigations were made in nature monuments, which are an important component of the nature protection system in Primorsky Territory, like nature reserves and national parcs (Seledets 2000b, 2004a, 2005a, 2008;

Syomkin et al. 2010). Coenopopulations of *A. hirta* are concentrated on the seacoasts; outside of these areas, only few are known, e.g., in the Senkina Shapka nature monument (near the town of Ussuryisk). Coenopopulations of *A. anomala* are typical for the Suifun-Khanka Plain, for the foothills and low mountain belt of Sikhote-Alin and the East Manchurian Mts, but sometimes they occur on the seacoasts and on the islands of Peter the Great Bay (the Sea of Japan).

The materials for our study were the descriptions of the typical plant communities with *Arundinella anomala* in various areas of Primorsky Territory, mostly in its intracontinental part. For comparison, typical seacoast plant communities with *A. hirta* were studied. The plant communities were chosen following the rule of maximal ecological variability of the habitats. Communities from the maritime belt, the coastal plains and the mountain slopes were selected. We also studied collections of the RFE Regional Herbarium (VLA) in the Institute of Biology and Soil Sciences FEB RAS (Vladivostok).

The comparative study of ERs and ecological niches in *Arundinellas* (Tables 3, 4, 5; Figure 13) allowed us to estimate the contribution of the coastal and submontane coenopopulations to the species ecological diversity. The ecological and phytocoenotic spectra of *Arundinella* spp. under study are generally similar, but some significant differences in their details emerged from the study of their ERs (Figure 13).

Table 4. Ecological characteristics of coenopopulations and floristic composition of plant communities with *Arundinella hirta* on the coast of Peter the Great Bay, the Sea of Japan (in grades of ecological scales)

Numbers of relevés groups	1	2	3	4	5	6	7	8	9	10
Humidity	68	66	70	63	67	65	65	64	64	66
Soil fertility and salinity	12	12	12	12	11	11	10	13	11	10
Granulometric composition of soils	4	8	5	5	8	5	3	3	3	3
Drainage	6	6	6	6	6	6	8	9	9	9
Recreational and pasture load	5	4	3	3	4	3	3	3	3	4
Variability of humidity	7	5	4	8	6	9	6	6	11	11
Renewal of the soil	12	13	12	8	9	7	11	11	7	4
Shading	2	2	2	2	4	7	5	4	2	2
Projective cover, %										
Artemisia gmelinii	-	-	-	-	-	-	-	-	-	+

Table 4. (Continued)

Numbers of relevés groups	1	2	3	4	5	6	7	8	9	10
A. littoricola	5	20	-	-	-	-	-	-	-	-
A. stolonifera	-	-	-	-	-	+	1	+	1	-
Arundinella hirta	+	10	20	5	+	70	80	80	70	1
Calamagrostis brachytricha	-	-	-	-	-	-	-	-	+	-
C. extremiorientalis	-	-	-	5	-	5	-	-	-	3
Carex bostrichostigma	-	-	-	-	-	-	+	-	-	-
C. siderosticta	-	-	-	-	-	-	1	-	-	-
C. ussuriensis	-	-	-	-	-	-	-	-	1	-
Cleistogenes kitagawae	-	-	-	-	-	-	-	+	10	-
Clinopodium chinense	-	-	-	+	1	+	-	1	-	-
Dianthus amurensis	-	-	-	+	+	+	-	-	+	-
Elymus gmelinii	-	-	-	-	3	-	-	-	-	-
E. woroschilowii	5	+	-	5	-	+	-	-	-	-
Euphrasia maximowiczii	-	-	-	-	-	+	-	-	+	-
Festuca ovina	-	-	-	-	-	+	-	-	+	-
Fragaria orientalis	-	-	-	-	-	-	1	1	+	-
Galium davuricum	-	-	-	-	-	+	1	1	-	-
G. verum	-	-	-	3	-	+	-	-	-	1
Geranium sibiricum	-	-	-	-	+	-	-	-	-	-
Glycine soja	-	-	-	-	-	+	10	1	-	-
Gypsophila pacifica	-	-	-	-	-	-	-	-	-	+
Hypericum attenuatum	-	-	-	-	+	+	-	-	-	-
Koeleria tokiensis	-	-	-	5	-	5	-	-	10	5
Leymus coreanus	-	-	+	+	+	+	+	+	+	+
L. mollis	-	-	-	-	-	-	-	-	-	5
Miscanthus sinensis s.l.	50	-	-	80	60	-	-	-	-	-
Orostachys malacophylla	+	+	-	-	-	-	-	-	-	-
Plantago depressa	-	-	-	-	-	+	-	-	-	+
Senecio cannabifolius	-	-	-	-	+	+	-	-	-	-
Setaria pumila	+	+	-	-	-	-	-	-	-	-
Spodiopogon sibiricus	-	-	-	-	-	-	-	-	5	-
Trifolium repens	-	-	+	-	3	-	-	-	-	-
Trisetum sibiricum	-	-	-	-	-	+	-	-	-	+
Vicia amoena	-	-	-	5	1	-	-	-	-	-

Table 5. Ecological niches of *Trisetum molle, Arundinella anomala, A. hirta, Agrostis trinii, Calamagrostis brachytricha, Leymus mollis* (in grades of ecological scales)

Trisetum molle

	A	B	C	D
H	62-88	70-100	65-80	60-90
SFS	3-6	3-9	5-10	6-18
Sh	1-4	1-4	1-4	1-4
G	6-12	6-14	3-9	1-12
D	8-11	3-7	4-8	2-11
VH	4-10	1-9	7-18	2-16
R	10-20	10-20	10-14	10-14
RPL	1-2	1-3	2-4	2-6

Arundinella anomala

H	55-65
SFS	7-8
Sh	1-10
G	2-12
D	7-12
VH	8-12
R	4-12
RPL	1-5

Arundinella hirta

H	65-70
SFS	8-9
Sh	2-6
G	3-10
D	6-8
VH	8-12
R	4-12
RPL	2-4

Agrostis trinii

	A	B
H	60-65	65-70
SFS	9-13	8-9
Sh	1-3	2-6
G	2-6	3-10
D	4-6	6-8
VH	5-11	8-12
R	8-14	4-12
RPL	3-6	2-4

Table 5. (Continued)

Calamagrostis brachytricha

	A	B	C
H	53-76	52-78	59-63
SFS	5-16	4-15	8-10
Sh	3-10	1-7	1-7
G	4-10	5-15	4-6
D	6-12	10-12	10-11
VH	3-15	5-15	9-15
R	3-19	1-7	3-7
RPL	1-5	1-4	1-3

Leymus mollis

	A	B
H	49-76	68-75
SFS	5-16	9-14
Sh	1-7	3-6
G	5-15	6-14
D	5-12	2-11
VH	1-11	1-12
R	3-18	2-20
RPL	1-6	1-5

Taking into account the combination of ecological factors, the dimensions of the ecological niche of *A. anomala* are considerably larger than those of *A. hirta*. We consider this fact evidence of the adaptation of *A. anomala* to a more widely varying spectrum of ecological factors (Table 5, Figure 13). Such an ecological spectrum is typical for plant species of intracontinental areas. *A. hirta* does not possess such a large ecological niche and is obviously not adapted to different habitats.

ERs are generally studied in terms of the most important factors, humidity (H) and soil fertility and salinity (SFS). This type of study showed that the ERs of the *Arundinella* species studied do not overlap, but only contact each another. As to adaptation of the species studied to SFS, *A. hirta* is tolerant to saline soils, but *A. anomala* avoids them. In the environmental situation of the RFE the adaptation to soil salinity means adaptation for marine habitats because saline soils are rarely encountered outside of the maritime belt except in a few cases connected with human activity.

Similar proportions of ecological factors in the ERs of *A. anomala* and *A. hirta* can be seen if the ERs are based on the combination of the humidity (H) and granulometric composition of the soil (G). The adaptation of *A. anomala* to G is from 2 to 12 degrees. *A. anomala* is adapted to the wide spectrum of G that occurs in intracontinental territories of Asia, but *A. hirta* is adopted mostly to the sandy, pebbly and rocky soils on the seacosts. Here, we can observe the edaphic border between *A. anomala* and *A. hirta*: *A. hirta* has the advantages for surviving on substrates of marine origin (pebbles, sand and other products of the seashore rock destruction, as well as undeveloped soils of seacosts), but *A. anomala* occurs on the soils of zonal type.

Ecological estimation of habitats of coenopopulations of *A. anomala* and *A. hirta* allows us to come to a conclusion concerning the degree of their specialization for environmental conditions in the RFE. The ecological niches of these species are similar in dimensions and configuration but differ significantly in the meaning and amplitudes of ecological factors. *A. hirta* has a higher maximal meaning of the principal ecological factors (H, SFS) than *A. anomala*. Thus, *A. hirta* is adapted to the seacoast environmental conditions with their high humidity and enrichment of soils by mineral nutrition. At the maritime belt and in adjacent areas, there is a high concentration of salt in the soil and in the air, especially during tides and storms.

The amplitude of changes in ecological factors is a reliable indicator of the specialization for certain types of habitats. The specialization of *A. hirta* is much more strict than that of *A. anomala*. It relates to the humidity, soil fertility and salinity. These ecological factors are especially typical for maritime vegetation. *A. anomala* tolerates a broad amplitude of change in the granulometric composition of soils, drainage and shading. These ecological factors are the most important in the intracontinental areas.

Thus, according to the combination of ecological characteristics, *A. hirta* is a typical coastal species, and *A. anomala* is a typical intracontinental species, although the latter can occur along seacosts too. On the seacoasts, *A. anomala* is not in its favorable ecological situation, rather it is in the marginal part of its ER.

The comparative study of ERs in *Arundinella* spp. allowed us to discover the peculiarities of their ecological differentiation. By the ERs with axes "humidity", "soil fertility and salinity", and "granulometric composition of soils", these species appear separated from one another, although they are contiguous. By the assemblage of ecological characteristics these species are also differentiated according to the oceanity of the climate: *A. anomala* is common in intracontinental areas, while *A. hirta* is common along the seacoasts. We suggest that the comparison of ERs and ecological niches as well as the geographical distributions of the two species prove these species distinct in this rank.

We consider *A. anomala* to be less specialized than *A. hirta* because it is confined to the vast area of the Pacific monsoon climate, whereas *A. hirta* is more specialized because it is confined to the narrow oceanic zone within this area.

CONCLUSION

The ecological range (ER) of a species is a part of its ecological niche; the niche reflects a species position in multidimensional ecological space.

The trends in the changes of the suite of indices of the ERs of species that occur in the transition from the continental climatic conditions of Inner Asia to the monsoon climate of the Pacific coast have been revealed. The ecological variability of the coenopopulations also depends on their position in the plant cover. In the geographical profile from Inner Asia to the Pacific monsoon zone, the properties of the ERs change. The relative position of the ecological optimum and the center of the ER as well as the distance between them show the direction of ecological adaptation of the taxa. As for the monsoon zone, considerable transformation of the ER occurred according to environmental conditions. Changes in the ER are indicators of the evolutionary processes. The continental climate causes the continentalization of ecological niches.

One application of the data obtained is a new, more differentiated approach to biodiversity conservation, with consideration for the positions of coenopopulations in the species area of distribution. Our studies show that it is necessary to protect not only results of evolution (coenopopulations, species and taxa of higher rank) but the conditions that lead to evolutionary processes in the specific situation of the monsoon zone, including the variety of habitats and natural processes in the contact zone. Nature protection should be carried out taking into consideration the structures of populations, the adaptive characteristics of the species, and their degree of specialization, as well as the ecological ranges of the species to be protected and their optima. Also the opportunities for a species to participate in evolutionary process must be evaluated.

REFERENCES

Akhtyamov, M. Kh. (1998). Monsoons and special features of meadows in south part of the Russian Far East. In: O. V. Khrapko, (Ed.), *Plants in monsoon climate*: Proceedings of the conference devoted to the 50-th anniversary of the Botanical Garden-Institute FEB RAS (3-5). Vladivostok: Dalnauka. (In Russian).

Baikalova, A. S. (2005). Changing of the state of coenopopulations of *Epipactis palustris* in Yugansky nature reserve. In: *Biodiversity and spatial organization of vegetation in Siberia, methods of study and protection*: Proceedings of All-Russian conference (Novossibirsk, 25-27 Oct. 2005). Novossibirsk, Central Siberian Botanical Garden, Siberian Branch of the Russian Academy of Sciences. (14-15). (In Russian).

Barkalov, V. Yu. (1987). *Boehmeria tricuspis* (Urticaceae) - new species of the USSR flora from the Kurile Islands. *Bot. Journ.*, (Leningrad), 72, 11, 1548-1550. (In Russian).

Barkalov, V. Yu. (1998). Vascular plants of the Kuril'sky nature reserve (Sakhalinskaya Oblast'). In: A.E. Kozhevnikov, (Ed.), *Flora of protected wildlife areas of the Russian Far East: Magadansky, Bureyinsky and Kuril'sky nature reserves* (71-113). Vladivostok: Dalnauka. (In Russian).

Barkalov, V. Yu. (2000). Phytogeography of the Kurile Islands. In: T. Komai, (Ed.), Results of recent research on Northeast Asian biota. *Nat. Hist. Res., Special Issue*, 7 (1-14). Chiba: Natural History Museum and Institute (Japan).

Barkalov, V. Yu. (2009). *Flora of the Kuril Islands*. Vladivostok: Dalnauka. (In Russian).

Barkalov, V. Yu. and Eremenko, N. A. (2003). *Flora of nature reserve "Kurilsky" and nature zakaznik "Malye Kurily"* (Sakhalinskaya Oblast). Vladivostok: Dalnauka. (In Russian).

Barkalov, V. Yu., Kozhevnikov, A. E. and Kharkevich, S. S. (1986).Vascular plants of Verkhoturova and Karaginsky Islands (the Bering Sea) and conservation of their gene pool. *Komarov Memorial Lectures* (Vladivostok), 33, 110-168. (In Russian).

Barkalov, V. Yu. and Yakubov, V. V. (2007). New species for the flora of Russia - *Psilotum nudum* (L.) Beauv. (Psilotaceae) from the Kurile Islands. *Bot. Zhurn.* (St.-Petersbourg), 92, 12, 154-156. (In Russian).

Bashtavoy, N. G. (1990). Coenopopulations of herbaceous plants in deciduous forests under conditions of anthropogenic load. In: *Botanical studies in Ukraine*. Kiev: Naukova Dumka, (6-7). (In Russian).

Bassargin, D. D. (2002). Diversity of population groups of *Saussurea pulchella* (Fisch.) Fisch. and *S. grandifolia* Maxim. (*Asteraceae*) on Muravyov-Amursky Peninsula (South Primorye). In: *Biological studies in Mountain-Taiga Station*. Collected papers, Issue 8. Vladivostok, (41-60). (In Russian).

Bassargin, D. D. and Vorobyova, A. N. (2003a). Problems of plant population-specific differentiation in the aspect of the Pacific effect. In: S. B. Gontcharova, (Ed.), Plants in monsoon climate, III: Proceedings of III International conference *"Plants in Monsoon Climate"* (198-199). Vladivostok: BGI FEB RAS. (In Russian).

Bassargin, D. D. and Vorobyova, A. N. (2003b). *Saussurea* (Asteraceae) species in the aspect of "Levins rule". In: S. B. Gontcharova, (Ed.), Plants in monsoon climate, III: Proceedings of III International conference *"Plants in Monsoon Climate"* (199-203). Vladivostok: BGI FEB RAS. (In Russian).

Bayer, R. J., Purdy, B. G. and Lebedyk, D. G. (1991). Niche differentiation among eight sexual species of *Antennaria* Gaertner (Asteraceae: Inuleae) and *A. rosea*, their allopolyploid derivative. *Evol. Trends in Plants*, 5, 109-132.

Bayer, R. J. (1998). New perspectives into the evolution of polyploid complexes. In: Plant evolution in man-made habitats. Proceedings of VII Symposium of IOPB, *Amsterdam* (359-373).

Bebbel, G. R. and Selander, R. K. (1975). Genetic variability in edaphically restricted and widespread plant species. *Evolution*, (USA), 28, 4, 619-630.

Bekmansurov M.V. and Zhukova L.A. (2000). *Indicative possibilities of plant species and ecological scales*. Field ecological practical studies. Yoshkar-Ola: Marij-El State University. (In Russian).

Binggeli P. (1996). A taxonomic, biogeographical and ecological overview of invasive woody plants. *Journ. Veget. Sci.*, 7, 1, 121-124.

Borzova, L. M., Klychkova, T. V., Probatova, N. S., Syomkin, B. I. and Kharkevich, S. S. (1985). The rediscovery of the rarest species *Dimeria neglecta* in the Primorsky Territory. *Bot. Journ.* (Leningrad), 70, 9, 1261-1265. (In Russian).

Box, E. O., You, Hai-Mei, Li, Dong-Liang. (2001). Climatic ultra-continentality and the abrupt boreal-nemoral forest boundary in northern Manchuria. Studies on the vegetation of alluvial plains (183-200).

Brochmann, C. and Elven, R. (1992). Ecological and genetic consequences of polyploidy in arctic *Draba* (Brassicaceae). *Evol. Trends in Plants*, 6, 111-124.

Bulokhov, A. D. (1996). Ecological estimation of habitats by methods of the phytoindication. Bryansk: the Bryansk State Pedagogical Univ. Press. 104 p. (In Russian).

Burda R. I. (1991). Anthropogenic transformation of the flora. Kiev: Naukova Dumka. (In Russian).

Burkovskaya, E. V., Burundukova, O. L. and Probatova, N. S. (2005). Structural and functional adaptations of the photosynthetic apparatus in coastal vascular plants - halophytes (southern Russian Far East). In: O. V. Khrapko, (Ed.), *Rhythms and catastrophes impact on the vegetation of the Russian Far East*: Proceedings of the International conference "Rhythms and catastrophes impact on the vegetation of the Russian Far East" (232-238). Vladivostok: BGI FEB RAS. (In Russian).

Burundukova, O. L., Neupokoyeva, E. V. and Probatova, N. S. (1997). Halophytes on seacoasts of southern Far East: anatomic and physiological aspects of adaptation. In: *Animal and plant life of the Far East*, 3, 210-215. Ussurijsk. (In Russian).

Bykov B. A. (1967). *Geobotanical terminology*. Alma-Ata: Nauka Publishing House of Khazakh SSR. (In Russian).

Cayouette, J. and Darbyshire, S. J. (1993). The intergeneric hybrid grass *"Poa labradorica"*. *Nord. Journ. Bot.* 13, 6, 615-629.

Cheremushkina, V. A. (2002). Ecological and phytocoenotic position of *Allium* L. representatives in mountain ecosystems of Eurasia. *Sib. Ecol. Journ.* 9, 5, 565-570. (In Russian).

Chiapella, J. and Probatova, N.S. (2003). The *Deschampsia cespitosa* complex (*Poaceae*: *Aveneae*) with special reference to Russia. *Bot. Journ. Linnean Soc.* (London), 142, 213-228.

Chubar, E. A. (1991). New species for the flora of the islands of the Far East marine reserve. *Bot. Zhurn.* (St.-Petersbourg), 76, 11, 1620-1622. (In Russian).

Chubar, E. A. (1992). Addition to the flora of the islands of the Far East marine reserve. *Bot. Zhurn.* (St.-Petersbourg), 77, 12, 131-133. (In Russian).

Chubar, E. A. (1994). Additions and changes to the flora of the islands of the Far East marine reserve. *Bot. Zhurn.* (St.-Petersbourg), 79, 7, 128-130. (In Russian).

Chubar, E. A. (1996). Additions to the flora of the islands of the Far East marine reserve. *Bot. Zhurn.* (St.-Petersbourg), 81, 10, 117-119. (In Russian).

Chubar, E. A. (1998). A finding of *Matteuccia orientalis* (*Onocleaceae*) in South Primorye and new species for the flora of the islands of the Far East marine reserve. *Bot. Zhurn.* (St.-Petersbourg), 83, 3, 150-152. (In Russian).

Chubar, E. A. (2008). *Cakile edentula* (Brassicaceae) – new genus and species for the flora of East Asia. *Bot. Zhurn.* (St.-Petersbourg), 93, 4, 634-637. (In Russian).

Chubar, E. A., Probatova, N. S. and Seledets, V. P. (2004). Vascular plants. In: A. N. Tyurin, and Drozdov, A. L. (Eds.), *Far Eastern marine biosphere reserve*, II, Biota (373-420). Vladivostok: Dalnauka. (In Russian).

Chumakova, E. A. (1988). Coenopopulation structure of ephemeroids in deciduous forests under conditions of recreational load. In: Relevant problems of botany in the USSR. Alma-Ata. (444-445).

Cody, M. L. (1991). Niche theory and plant growth forms. *Vegetatio*, 97, 1, 39-55.

Coenopopulations of plants: basic notions and structure. (1976). Moscow: Nauka. (In Russian).

Coenopopulations of plants: development and relationships. (1977). Moscow: Nauka. (In Russian).

Coenopopulations of plants: outlines of population biology. (1988). Moscow: Nauka. (In Russian).

Darbyshire, S. J., Cayouette, J., and Warwick, S. I. (1992). The intergeneric hybrid origin of *Poa labradorica (Poaceae)*. *Pl. Syst. Evol.* 181, 57-76.

Dynamics of plant coenopopulations. (1985). Moscow: Nauka.

Ehrendorfer, F. (1980). Polyploidy and distribution. In: W. H. Lewis (Ed.), *Polyploidy: Biological Relevance* (45-60). New York.

Flora of the Russian Far East: Alphabetical Indexes to "Vascular Plants of the soviet Far East", vols. 1-8 (1985-1996). Kozhevnikov, A. E. and Probatova, N. S. (Eds). (2002). Vladivostok: Dalnauka. (In Russian).

Flora of the Russian Far East: Addenda and corrigenda to "Vascular plants of the Soviet Far East", vols. 1-8 (1985-1996). Kozhevnikov, A. E. and Probatova, N. S. (Eds.). (2006). Vladivostok: Dalnauka. (In Russian).

Gabeev, M. D., Dutikova, V. A., Kartsev, G. A., Kulikov, E. P., Shilov M. P., and Shorokhova, M. V. (1973). *Provisional instructions for geobotanical investigation of the nature fodder grounds for collective and state farms and other agricultural enterprises of the RSFSR.* Moscow: Rosgiprozem Press.

Galanin, A. V. and Belikovich, A. V. (2009). The Pacific monsoon zone: phytogeographical regionalization, plant migrations, and distinctive features of speciation. In: A. V. Belikovich, (Ed.), *Plants in Monsoon Climate V*: Proceedings of V Conference "Plants in Monsoon Climate" (33-43). Vladivostok: Dalnauka. (In Russian).

Giller, P. S. (1984). Community structure and the niche. London.

Gleason, H. A. (1917). The structure and development of the plant associations. *Bull. Torrey Bot. Club*, 44, 10, 463-481.

Gleason, H. A. (1924). The individualistic concept of the plant association. *Bull. Torrey Bot. Club,* 53, 7-26.

Gorchakovsky, P. L. (1979). The trends of anthropogenic changes of vegetation of the Earth. *Bot. Journ.* (Leningrad), 64, 12, 1697-1714. (In Russian).

Gorovoy, P. G., Chubar, E. A., andVolkova, S. A. (1999). New genus *Limonium* (*Limoniaceae*) in the flora of the Russian Far East and a new species *L. tetragonum* in the flora of Russia. *Bot. Journ.* (St.-Petersbourg), 84, 7, 144-147. (In Russian).

Grant, V. (1963). *The origin of adaptations.* New York.

Gray, A. J., Parsell, R. J., and Scott, R. (1979). The genetic structure of plant populations in relation to the development of salt marshes. In: *Ecological processes in coastal environment.* First Ecol. Symp. and 19-th Symp. Brit. Soc., Norwich, 1977. (43-64). Oxford.

Guilarov, A. M. (1978). Contemporary state of the concept of ecological niche. *Progress in contemporary biology*, 85, 3, 431-446. (In Russian).

Guilarov, A. M. (1990). Population ecology. Moscow: The Moscow Univ. Press.

Hanski, I. (1998). Metapopulation dynamics. *Nature*, 396, 41-49.

Hanski, I. (1999). *Metapopulation ecology.* Oxford.

Harper, L. (1977). Population biology of plants. London; New York: Acad. Press.

Hutchinson, G. E. (1965). The niche: an abstractly inhabited hyper-volume. In: *The ecological theatre and the evolutionary play*. New Haven: Yale University Press. (26-78).

Hutchinson G. E. (1978). *An introduction to population ecology*. New Haven and London: Yale University Press.

Ipatov, V. S., and Kirikova, L. A. (1997). *Phytocoenology*: Handbook. St.-Petersbourgh: the St.-Petersbourgh State University Publishing House. (In Russian).

Ipatov, V. S. and Kirikova, L. A. (2001). Reactivity and sensitivity of species to ecological factors. *Bot. Journ.* (St.-Petersbourg), 86, 3, 80-86. (In Russian).

Kandalova, G. T. (2005). Positions of the coenopopulation edificators in various regimes of maintenance of restoring meadow steppes in Khakassia (on the example of the nature reserve "Khakassky"). In: *Biodiversity and spatial organization of vegetation in Siberia, methods of study and protection*: Proceedings of All-Russian conference (Novossibirsk, 25-27 Oct. 2005). Novossibirsk, Central Siberian Botanical Garden, Siberian Branch of the Russian Academy of Sciences. (73-75). (In Russian).

Khristoforova, N. K. (1999). *Fundamentals of Ecology*: Handbook for biological and ecological departments of universities. Vladivostok: Dalnauka. (In Russian).

Kitayama, K. and Mueller-Dombois, D. (1995). Biological invasion on an oceanic island mountain: do alien plant species have wider ecological ranges than native species? *Journ. Veget. Sci.*, 6, 5, 667-674.

Komarov, V. L. (1937). Vegetation of seacoasts in Kamchatka Peninsula. *Transactions of the Far East Branch of USSR Academy of Sciences*, Ser. Bot., 2, 7-17. (In Russian).

Komarova, T. A. (1992a). Afterfire successions in forests of the south Sikhote-Alin. Vladivostok: Far East Branch of the USSR Academy of Sciences. (In Russian).

Komarova, T. A. (1992b). Development and productivity of herbaceous and shrub coenopopulations (forests of the south Sikhote-Alin). Vladivostok: Far East Branch of the USSR Academy of Sciences. (In Russian).

Komarova, T. A., Atschepkova, L. Ya. (2000). Working out of regional ecological scales and their using for classification of the forests with *Pinus koraiensis*. *Komarov Memorial Lectures* (Vladivostok), Issue 46. Vladivostok: Dalnauka, (7-72). (In Russian).

Komarova, T. A., Timoszenkova, E. V., Prokhorenko, N. B., Aszepkova, L. Y., Yakovleva, A. N., Sudakov, Yu. N. and Seledets, V. P. (2003). *Regional ecological scales and their application for classification of the forest vegetation of the Russian Far East*. Vladivostok: Dalnauka. (In Russian).

Kozhevnikov, A. E. (1985). *Carex tegulata (Cyperaceae)* - a species new for the flora of the USSR. *Bot. Journ.* (Leningrad), 70, 2, 272-274. (In Russian).

Kozhevnikov, A. E. (2001). *The Sedge Family (Cyperaceae Juss.) of the Russian Far East* (taxonomic composition and main regularities of its formation). Vladivostok: Dalnauka.

Kozhevnikov, A. E. , and Korkishko, R. I. (1985). *Carex holotricha* Ohwi - a species new for the flora of USSR. *Bulletin of the Main Botanical Garden of Academy of Sciences of the USSR*, Issue 135 (32-36). (In Russian).

Kozhevnikov, A. E. , and Kozhevnikova, Z. V. (2000a). A finding of *Mitrasacme indica* on the Russian Far East, the Family Loganiaceae, new for the flora of Russia. *Bot. Journ.* (St.-Peterbourg), 85, 5, 130-134. (In Russian).

Kozhevnikov, A. E., and Kozhevnikova, Z. V. (2000b). The genus *Lipocarpha (Cyperaceae)*, new for the flora of Russia, found on the Russian Far East (the Primorsky Territory). *Bulletin of the Moscow society of nature explorers*, Biol., 105, 2, 58. (In Russian).

Kozhevnikov, A. E., and Kozhevnikova, Z. V. (2000c). *Utricularia caerulea* L. *(Lentibulariaceae)* - a species new for the flora of Russia, from the Khassansky nature parc (the Primorsky Territory). *Bulletin of the Moscow society of nature explorers*, Biol., 105, 3, 66-68. (In Russian).

Kozhevnikov, A. E., and Kozhevnikova, Z. V. (2001a). *Hypericum laxum (Hypericaceae)* - a species new for the flora of Russia (the Primorsky Territory). *Bot. Journ.* (St.-Peterbourg), 86, 4, 160-163. (In Russian).

Kozhevnikov, A. E., and Kozhevnikova, Z. V. (2001b). New findings of the warm-temperate and subtropical relict elements of the flora on the southwest of the Primorsky Territory. In: *Biological studies in the Mountain-Taiga Station*, (Collected papers), Issue 7, Vladivostok (188-193). (In Russian).

Kozhevnikov, A. E. and Kozhevnikova, Z. V. (2007). Flora of the Amur River Basin (the Russian Far East): taxonomic diversity and spatial changings of taxonomic structure. *Komarov Memorial Lectures* (Vladivostok), Issue 55. Vladivostok: Dalnauka, (104-183). (In Russian).

Kozhevnikov, A. E. and Kozhevnikova, Z. V. (2008). On the taxonomy of the species of *Ophelia* D. Don *(Gentianaceae* Dumort.*)* in the flora of Russia. *Bulletin of the Moscow society of nature explorers*, Biol., 113, 5, 66-68. (In Russian).

Kozhevnikov, A. E., Sokolovskaya, A. P., and Probatova, N. S. (1986). Ecology, geography and chromosome numbers of some *Cyperaceae* species from the soviet Far East. *Proceedings of the Siberian Branch of the USSR Academy of Sciences*. Biol., 13, 2, 57-62. (In Russian).

Kozhevnikova, Z. V. (2003). *Viola diamantiaca* Nakai on the Russian Far East. *Turczaninowia*, 6, 4, 27-34. (In Russian).

Kravchunovskaya, E. A., Myalo, E. G. and Sidneva, E. N. (2009). Plant communities of multiple-aged marine terraces in Avachinsky Bay, East Kamchatka. In: A. V. Belikovich, (Ed.), *Plants in the Monsoon Climate. V: Proceedings of V Conference "Plants in Monsoon Climate"* (88-90). Vladivostok: Dalnauka. (In Russian).

Kravchunovskaya, E. A., Pinegina, T. K. and Sergueyeva, A. M. (2008). Plant communities of the multiple-aged coastal banks in Kamchatka. In: Proceedings of the IX International scientific conference *"Conservation of biodiversity of Kamchatka and adjacent seas"*. Petropavlovsk-Kamchatsky, (70-73). (In Russian).

Kurentsova, G. E. (1952). Grasslands with *Arundinella* in the east part of the Khanka Plain, the Primorsky Territory. *Transactions of the Far East Branch of the USSR Academy of Sciences*, Issue 4 (23-27). (In Russian).

Kurentsova, G. E. (1955). The origin of vegetation of the Khanka Plain. *Bot. Journ.* (Leningrad), 40, 2, 178-188. (In Russian).

Kussakin, O. G., Kubanin, A. A., Bregman, Yu. E., et al. (1993). Fishery and coastal complex. In: *Long-term program of nature protection and rational use of the Primorsky Territory nature resources till 2005* (Ecological program), Vladivostok: Dalnauka, Part 1, (229-286). (In Russian).

MacArthur, R. H. and Wilson, E. O. (1967). *The theory on island biogeography*. Princeton: Princeton Univ. Press.

Malyshev, L. I. (1993). Ecological background of the floristic diversity in Northern Asia. *Fragm. Florist. Geobot.*, Suppl. 2 (1), 331-342.

McIntosh, R. (1975). H. A. Gleason "individualistic ecologist" 1882-1975. His contribution to ecological theory. *Bull. Torrey Bot. Club,* 102, 263-273.

McIntosh, R. P. (1983). Excerpts from the works of L. G. Ramensky. *Bull. of the Ecol. Soc. of America*, 64, 7-12.

Matyukhin, D. L. and Manina, O. S. (2007). Structural peculiarities of shoots in the species and varieties of *Juniperus* L. originated from the regions with monsoon climate. In: S. B. Gontcharova, (Ed.), *Plants in monsoon climate*, 2006. IV. Proceedings of IV International conference "Plants in Monsoon Climate" (383-384). Vladivostok: BGI FEB RAS. (In Russian).

Mayorov, I. S., Seledets, V. P. and Probatova, N. S. (2009). Ecoranges of plant species in adaptive zones. *Herald of the Krasnoyarsk State University of Agriculture*, 28, 2, 47-56. (In Russian).

Mirkin, B. M. (1984). Anthropogenic dynamics of vegetation. In: *Results of Science and Technology*, Ser. Botany, 5, Geobotany. Moscow: VINITI, (139-232). (In Russian).

Mirkin, B. M. (1987). Paradigm change and the vegetation classification in Soviet phytocoenology. *Vegetatio*, 68, 131-138.

Mirkin, B. M. and Naumova, L. G. (1998). *Vegetation science* (history and actual state of principal concepts). Ufa: Gilem. (In Russian).

Mirkin, B. M., Naumova, L. G., and Solometsch, A. I. (2000). *Modern vegetation science*. Handbook. Moscow: Logos. (In Russian).

Mirkin, B. M. and Solometch, A. I. (1989). Syntaxonomy of synanthropic vegetation: recent state and the trends of development. *Journ. Fundam. Biol.*, 50, 3, 379-387. (In Russian).

Monsoon Climate Plants. (2000). II International conference *"Monsoon Climate Plants"*. (Abstr.). O. V. Khrapko (Ed.). Vladivostok: Dalnauka. (In Russian).

Moravec, J. (1989). Influences of the individualistic concept on vegetation syntaxonomy. *Vegetatio*, 83, 29-39.

Morozov, V. L. and Belaya, G. A. (1998a). Ecological and phytocoenotic positions of steppe meadow plant communities in Primorye and Priamurye. In: O. V. Khrapko, (Ed.), *Plants in monsoon climate*. Proceedings of the conference devoted to the 50-th anniversary of the Botanical Garden-Institute FEB RAS (41-43). Vladivostok: Dalnauka. (In Russian).

Morozov, V. L. and Belaya, G. A. (1998b). Principles of phytogene pool and coenotic pool protecting of the most important terrestrial ecosystems in the North-East Asia. In: O. V. Khrapko, (Ed.), *Plants in monsoon climate*. Proceedings of the conference devoted to the 50-th anniversary of the Botanical Garden-Institute FEB RAS (268-270). Vladivostok: Dalnauka. (In Russian).

Morozov, V. L. and Belaya, G. A. (2009). Ecological specificity of the meadow herbs in steppe-like communities in the Primorsky Territory and the Amur River basin. In: A. V. Belikovich, (Ed.), *Plants in the Monsoon Climate*. V. Proceedings of V Conference "Plants in Monsoon Climate" (99- 101). Vladivostok: Dalnauka. (In Russian).

Nechaev, A. P. and Gapeka, Z. I. (1970). Ephemeral plants of the low-water season belt of riversides on the Lower Amur River. *Bot. Journ.,* (Leningrad), 55, 8, 1127-1137. (In Russian).

Neshatayeva, V. Yu. (1988). Vegetation of coastal meadows in East Kamchatka (on the example of Kronotsky state reserve). *Herald of Leningrad State University. Ser.* 3, *Issue* 4, 24, 35-44. (In Russian).

Neupokoyeva, E. B., Burundukova, O. L., and Probatova, N. S. (1998). Types of photosynthesis, structural and functional properties of sea coastal halophytes in the south of the Russian Far East. In: O. V. Khrapko, (Ed.), *Plants in monsoon climate.* Proceedings of the International conference devoted to the 50-th anniversary of the Botanical Garden-Institute FEB RAS (126-128). Vladivostok: Dalnauka. (In Russian).

Palmer, M. W. and van der Maarel, E. (1995). Variance in species richness, species association and niche limitation. *Oikos,* 73, 203-213.

Parrish, J. A. and Bazzaz, F. A. (1976). Underground niche separation in successional plants. *Ecology,* 57, 6, 1281-1288.

Patrievskaya, G. F. (1959a). Characteristics of "xerophytes" from *Arundinella* formation in the Khanka Plain. *Bot. Journ.* (Leningrad)*,* 44, 11, 1578-1592. (In Russian).

Patrievskaya, G. F. (1959b). Materials for characteristics of *Arundinella* meadows in Khanka Plain. *Proceedings of the Siberian Branch of the USSR Academy of Sciences*, 5. (112-120) (In Russian).

Pavlova, N. S. (1999). Floristic findings on the Slannikova Mt. (Sakhalin). *Bot. Journ.* (St.-Petersbourg), 84, 3, 129-133. (In Russian).

Pavlova, N. S. (2000). New findings of the rarest species of *Iris* L. in the Primorsky Territory. In: *Monsoon Climate Plants,* II International conference "Monsoon Climate Plants". (Abstr.). O. V. Khrapko (Ed.). Vladivostok: Dalnauka, (152-153). (In Russian).

Pavlova, N. S. (2006). On the finding of *Limonium tetragonum (Limoniaceae)* in continental part of the Primorsky Territory. *Bot. Journ.* (St.-Petersbourg), 91, 9, 1420-1423. (In Russian).

Pavlova, N. S. and Gorovoy, P. G. (1971). *Achyrophorus crepidioides* (Miyabe et Kudo) Kitag. - a species new for the flora of the USSR. *Bulletin of the Main Botanical Garden of Academy of Sciences of the USSR*, Issue 80 (21-24). (In Russian).

Pavlova, N. S., Probatova, N. S., Sokolovskaya A. P. (1989). Taxonomic review of the Family *Fabaceae*, chromosome numbers and distribution on the soviet Far East. Komarov Memorial Lectures (Vladivostok), Issue 36 (20-36). (In Russian).

Petukhova, I. P. (2003). Importance of drought resistance of plants in conditions of monsoon climate of the southern Primorye. In: S. B. Gontcharova, (Ed.), *Plants in monsoon climate*. III: Proceedings of III International conference "Plants in Monsoon Climate" (399-402). Vladivostok: BGI FEB RAS. (In Russian).

Plants in monsoon climate. (1998). *Proceedings of the conference devoted to the 50-th anniversary of the Botanical Garden-Institute* FEB RAS . O. V. Khrapko (Ed.). Vladivostok: Dalnauka. (In Russian).

Plants in monsoon climate. (2003). III. Proceedings of III International conference "*Plants in Monsoon Climate*". S. B. Gontcharova (Ed.). Vladivostok: BGI FEB RAS. (In Russian).

Plants in monsoon climate. (2007). IV. Proceedings of IV International conference "*Plants in Monsoon Climate*". S. B. Gontcharova (Ed.). Vladivostok: BGI FEB RAS. (In Russian).

Plants in the Monsoon Climate. (2009). V. Proceedings of V Conference "*Plants in Monsoon Climate*". A. V. Belikovich (Ed.). Vladivostok: Dalnauka. (In Russian).

Plotnikova, L. S. and Trulevich, N. V. (1974). Vegetation in the middle part of Karaguinsky Island west coast . In: P. I. Lapin, (Ed.). *Phyto-geographical regions of the USSR*. Prospects of plant introduction. (36-42). Moscow: Nauka (In Russian).

Prilutzky, A. N. (2007). Philosophical, conceptual and methodological problems in contemporary phytocoenology. In: S. B. Gontcharova, (Ed.), *Plants in monsoon climate*, 2006. IV. Proceedings of IV International conference "Plants in Monsoon Climate" (6-11). Vladivostok: BGI FEB RAS. (In Russian).

Probatova, N. S. (1961). Flood influence on the aquatic and riverside vegetation of the floodplain lakes in the Amur River basin. *Papers of the XIV scientific student conference*. Rostov-on-Don: the Rostov State University, (81-84). (In Russian).

Probatova, N. S. (1965). On biology of *Polygonum nodosum* under conditions of the Amur River floodplain. *The 8-th Conference of young scientists of the Far East,* Sect. Biology, Vladivostok: Institute of Biology and Soil Science, (43-46). (In Russian).

Probatova, N. S. (1969a). Adaptation of *Polygonum nodosum* Pers. to flood under conditions of the Amur River floodplain. *Bot. Journ.* (Leningrad), 54, 5, 755-760. (In Russian).

Probatova, N. S. (1969b). On the taxonomic position of *Poa radula* Franch. et Savat. и *P. ussuriensis* Roshev. In: *Problems in botany on the Far East* (to the 100-th anniversary of Academician V. L. Komarov). Vladivostok., (93-105). (In Russian).

Probatova, N. S. (1970). *Glyceria depauperata* Ohwi - new species in the flora of the USSR. *Bot. Journ.* (Leningrad), 55, 6, 876-877. (In Russian).

Probatova, N. S. (1971). New species of *Poa* L. from the Far East. *Novitates systematicae plantarum vascularium*, 8, 25-57. Leningrad: Nauka. (In Russian).

Probatova, N. S. (1973). New and rare species of *Poaceae* from the Far East. *Novitates systematicae plantarum vascularium*, 10, 68-79. Leningrad: Nauka. (In Russian).

Probatova, N. S. (1974a). *Gramineae*. In: A. I. Tolmachev, (Ed.), Key for the vascular plants of Sakhalin and Kurile Islands (56-89). Leningrad: Nauka. (In Russian).

Probatova, N. S. (1974b). On the new genus *Arctopoa* (Griseb.) Probat. *(Poaceae). Novitates systematicae plantarum vascularium*, 11, 44-54. Leningrad: Nauka. (In Russian).

Probatova, N. S. (1975a). *Poa shumushiensis, P. sachalinensis, P. macrocalyx, P. pseudoattenuata, P. ussuriensis.* In: *List of plants from Herbarium of the flora of USSR*, 20, Issue 110, numbers 5453-5457, Leningrad: Nauka, (64-66). (In Russian).

Probatova, N. S. (1975b). New information about *Arctopoa trautvetteri* (Tzvel.) Probat. (*Poaceae*). *Novitates systematicae plantarum vascularium*, 12, 9-11. Leningrad: Nauka. (In Russian).

Probatova, N. S. (1976). New and rare species of *Poaceae* from the East Siberia and the Far East. *Novitates systematicae plantarum vascularium*, 13, 32-42. Leningrad: Nauka. (In Russian).

Probatova, N. S. (1979a). On some *Poaceae* species from the Far East. *Novitates systematicae plantarum vascularium*, 1978, 15, 68-75. Leningrad: Nauka. (In Russian).

Probatova, N. S. (1979b). Particular features of taxonomic composition, the distribution and caryological characteristics of *Poaceae* in the mountain regions of Kamchatka Peninsula. *Problems in Botany*, 14, 1, 29-33. (In Russian).

Probatova, N. S. (1979c). *Agrostis alaskana, A. exarata, Arctopoa eminens, A. subfastigiata, A. trautvetteri, Calamagrostis distantiflora, C. sachalinensis* subsp. *litwinowii, Poa almasovii, P. sichotensis, P. skvortzovii*. In: *List of plants from Herbarium of the flora of USSR*, 22, Issue 117, numbers 5804-5813, Leningrad: Nauka, (46-51). (In Russian).

Probatova, N. S. (1979d). The genus *Poa* L. in the North Pacific area. *XIV Pacific science congress* (Khabarovsk, 1979), Committee H - Botany. Abstr. Moscow (86-87).

Probatova, N. S. (1981). Family *Poaceae* (*Gramineae*). In: S. S. Kharkevich and S. K. Cherepanov (Eds.), *Manual of the vascular plants of Kamchatskaya Oblast*. (332-370). Moscow: Nauka. (In Russian).

Probatova, N. S. (1984). New taxa of *Poaceae* from the Far East of the USSR. *Bot. Journ.* (Leningrad), 69, 2, 251-259. (In Russian).

Probatova, N. S. (1985). *Poaceae* Barnh. In: S. S. Kharkevich and N. N. Tzvelev (Eds.), Vascular plants of the soviet Far East, 1 (89-382). Leningrad: Nauka. (In Russian).

Probatova, N. S. (1989). Caryotaxonomic analyzis of the *Poaceae* in Pacific regions of the North Asia. *The 2-d Conference on plant caryology* (Novossibirsk, 1989). Abstr. of papers (29-31). (In Russian).

Probatova, N. S. (1993). The review of the Family *Lamiaceae* in the flora of the Russian Far East. *Komarov Memorial Lectures* (Vladivostok), Issue 41, Vladivostok: Dalnauka, (29-53). (In Russian).

Probatova, N. S. (1995a). The genus *Arctopoa* (Griseb.) Probat. *(Poaceae)*: Siberian - North Pacific contacts. In: *Problems in studies of the Siberian plant cover*. Abstr. of the conference (51-53). Tomsk: Tomsk University Press (In Russian).

Probatova, N. S. (1995b). *The North Pacific area: a zone of speciation for Poaceae*. In: XVIII Pacific Science Congress. Abstr. (192). Beijing, China.

Probatova, N. S. (1997a). *On the problem of the grass flora genesis in the N Pacific islands: the Commander Islands*. In: P. Ya. Baklanov and L. D. Filatova (Eds.), *Geographical studies in the Far East*. Proceedings of the conference devoted to the 150-th anniversary of the Russian geographical society (149-151). Vladivostok: Dalnauka. (In Russian).

Probatova, N. S. (1997b). *Taxonomic problems in biodiversity of the thick-rhizomatous sweet-grasses (Hierochloë R. Br., Poaceae) in the North-East Asia.* In: P. Ya. Baklanov and L. D. Filatova (Eds.), *Geographical studies in the Far East.* Proceedings of the conference devoted to the 150-th anniversary of the Russian geographical society (151-153). Vladivostok: Dalnauka. (In Russian).

Probatova, N. S. (1998). Caryology of the vascular flora in the south of the Russian Far East: analytical aspect. In: O. V. Khrapko, (Ed.), *Plants in monsoon climate.* 1998. Proceedings of the Conference devoted to the 50-th anniversary of the Botanical Garden-Institute FEB RAS (132-134). Vladivostok: Dalnauka. (In Russian).

Probatova, N. S. (2000). Caryology and the nature protection problems of the bank flora in the temperate regions of the monsoon climate. In: O. V. Khrapko, (Ed.), Monsoon Climate Plants. 2000. II Intern. conference *"Monsoon Climate Plants".* (Abstr.). (163-164). Vladivostok: Dalnauka. (In Russian).

Probatova, N. S. (2003a). Family *Poaceae.* In: S. S. Kharkevich and N. N. Tzvelev (Eds.). *Vascular Plants of the Russian Far East. Vol. 1. Lycopodiophyta, Juncaceae, Poaceae (Gramineae)* (87-488). Science Publishers Inc., Enfield, New Hampshire, USA.

Probatova, N. S. (2003b). The genus *Arctopoa* (Griseb.) Probat. (*Poaceae*): taxonomy, chromosome numbers, biogeography, differentiation. *Komarov Memorial Lectures* (Vladivostok), 49, 89-130. (In Russian).

Probatova, N. S. (2003c). Chromosome numbers of plant species as a source of information in studies on the flora of the Russian Far East. *Herald of the Far East Branch of the Russian Academy of Sciences* (Vladivostok), 3, 54-67. (In Russian).

Probatova, N. S. (2004). Specific features of the Russian Far East representatives of the Family *Poaceae*, in comparison with Siberia. In: *Problems of conservation of the Inner Asia vegetation*, I. Proceedings of the scientific conference (169-171). Ulan-Ude: Buryat Scientific Center SB RAS Publishers. (In Russian).

Probatova, N. S. (2006). *Poaceae.* In: A. E. Kozhevnikov, and N. S. Probatova (Eds.), Flora of the Russian Far East. *Addenda and corrigenda to "Vascular plants of the Soviet Far East"*, vols. 1-8 (1985-1996) (327-391). Vladivostok: Dalnauka. (In Russian).

Probatova, N. S. (2007). Chromosome numbers in the Family *Poaceae* and their significance for systematics, phylogeny, and phytogeography (the Russian Far East). *Komarov Memorial Lectures* (Vladivostok), 55, 9-103. (In Russian).

Probatova, N. S. (2008). Caryological study of *Poaceae* in the contact land-ocean zone (Russian Far East). In: M. E. Ermokhin, (Ed.), *Problems of marginal structures of biocoenoses*. Proceedings of the 2-d All-Russia scientific conference with international participation (206-210). Saratov: Saratov State University Press. (In Russian).

Probatova, N. S. and Barkalov, V. Yu. (2003). The Grass Family (*Poaceae*) on Sakhalin and the Kuril Islands: a comparative study of taxonomy and distribution. In: *Phytogeography of Northeast Asia: tasks for the 21-st century*. Abstr. of the Symposium (73). Vladivostok.

Probatova, N. S., Barkalov, V. Yu. and Rudyka, E. G. (2004). Chromosome numbers of selected vascular plant species from Sakhalin, Moneron and the Kurile Islands. In: H. Takahashi and M. Ōhara (Eds.), *Biodiversity and Biogeography of the Kuril Islands and Sakhalin,* 1. (Bulletin of the Hokkaido University Museum, 2) (15-23). Hokkaido University Museum, Sapporo, Japan.

Probatova, N. S., Barkalov, V. Yu. and Rudyka, E. G. (2007). *Caryology of the flora of Sakhalin and the Kurile Islands. Chromosome numbers, taxonomic and phytogeographical comments.* Vladivostok: Dalnauka. (In Russian, with English parts).

Probatova, N. S., Barkalov, V. Yu., Rudyka, E. G. and Kozhevnikova, Z. V. (2009). Additions to chromosome numbers for vascular plants from Sakhalin and the Kurile Islands (1). In: H. Takahashi and M. Ōhara (Eds.), *Biodiversity and Biogeography of the Kuril Islands and Sakhalin,* 3. (Bulletin of the Hokkaido University Museum, n° 5) (35-47). Sapporo: Hokkaido University Museum (Japan).

Probatova, N. S., Barkalov, V. Yu., Rudyka, E. G. and Pavlova, N. S. (2006). Further chromosome studies on vascular plant species from Sakhalin, Moneron and Kurile Islands. In: H. Takahashi and M. Ōhara (Eds.), *Biodiversity and Biogeography of the Kuril Islands and Sakhalin,* 2. (Bulletin of the Hokkaido University Museum, n° 3) (93-110). Sapporo: Hokkaido University Museum (Japan).

Probatova, N. S., Barkalov, V. Yu., Rudyka, E. G. and Shatalova, S. A. (2000). Chromosome study on vascular plants of the Kurile islands. In: T. Komai, (Ed.), Results of recent research on Northeast Asian biota. *Nat. Hist. Res., Special Issue,* 7 (21-38). Chiba: Natural History Museum and Institute (Japan).

Probatova, N. S., Bezdeleva, T. A. and Rudyka, E. G. (2001). Chromosome numbers, taxonomy and geographical distribution of the Far Eastern *Viola* (*Violaceae*). *Komarov Memorial Lectures* (Vladivostok), 48, 85-124. (In Russian).

Probatova, N. S. and Buch, T. G. (1981). *Hydrilla verticillata* (L. fil.) Royle (*Hydrocharitaceae*) in the Soviet Far East. *Bot. Journ.,* (Leningrad), 66, 2, 208-214. (In Russian).

Probatova, N. S. and Kharkevich, S. S. (1983). New taxa of *Poaceae* from the Khabarovsky Territory. *Bot. Journ.,* (Leningrad), 68, 10, 1408-1414. (In Russian).

Probatova, N. S., Korobkov, A. A., Gnutikov, A. A., Rudyka, E. G., Kotseruba V. V., and Seledets V. P. (2010). IAPT/IOPB chromosome data 10 / ed. by Karol Marhold. *Taxon,* 59, 6, 1935-1937, E6-E10.

Probatova, N. S., Kozhevnikova, Z. V., Rudyka, E. G., Kozhevnikov, A. E. and Nechaev V. A. (2010). Chromosome numbers of some vascular plants from the Far East of Russia. *Bot. Journ.,* (Leningrad), 95, 7, 1010-1022. (In Russian).

Probatova, N. S. and Olonova, M. V. (1991). The genus *Poa* L. in Siberia and the Far East of the USSR: comparative study of taxonomy and distribution. In: *Systematics and evolution in the Grass Family.* Krasnodar: the Kubansky State University, (96-98). (In Russian).

Probatova, N. S. and Rudyka, E. G. (1981). Chromosome numbers in some species of vascular plants of the Far East. *Proceedings of the Siberian Branch of the USSR Academy of Sciences.* Biol., 10, 2, 77-82. (In Russian).

Probatova, N. S. and Rudyka, E. G. (2000). Caryological study on the flora of Russky Island in Peter the Great Bay (the Primorsky Territory). In: O. V. Khrapko, (Ed.), Monsoon Climate Plants, 2000. II International conference *"Monsoon Climate Plants".* (Abstr.). (165-166). Vladivostok: Dalnauka. (In Russian).

Probatova, N. S. and Rudyka, E. G. (2003). Chromosome numbers in vascular flora of the Russian Far East: contemporary stage of the study. In: Botanical studies in Asiatic Russia, 1. Proceedings of XI-th Meeting of the Russian Botanical Society (311-312). Barnaul: AzBuka Publishing House. (In Russian).

Probatova, N. S., Rudyka, E. G. and Barkalov, V. Yu. (2003). Caryological studies on the vascular flora of sea coasts and islands of the Russian Far East. In: S. B. Gontcharova, (Ed.), *Plants in monsoon climate.* III. Proceedings of the 3-d International conference *"Plants in monsoon climate"* (283-288). Vladivostok: BGI FEB RAS. (In Russian).

Probatova, N. S., Rudyka, E. G., Kozhevnikov, A. E. and Kozhevnikova, Z. V. (2004). Chromosome numbers in representatives of the flora of the Primorsky Territory. *Bot. Journ.,* (St.-Petersbourg), 89, 7, 1208-1216. (In Russian).

Probatova, N. S., Rudyka, E. G. and Shatalova, S. A. (2001). Chromosome numbers in some species of the flora in outskirts of Vladivostok city (the Primorsky Territory). *Bot. Journ.,* (St.-Petersbourg), 86, 1, 169-172. (In Russian).

Probatova, N. S., Rudyka, E. G. and Shatokhina, A. V. (2007). Advance in chromosome numbers study on vascular flora of the Russian Far East in 2000-2006. In: S. B. Gontcharova, (Ed.), *Plants in monsoon climate*, 2006, IV. Proceedings of IV International conference "Plants in Monsoon Climate" (11-29). Vladivostok: BGI FEB RAS. (In Russian).

Probatova, N. S., Rudyka, E. G. and Sokolovskaya, A. P. (1998). Chromosome numbers in vascular plants from the islands of Peter the Great Bay and Muravyev-Amursky Peninsula (the Primorsky Territory). *Bot. Journ.,* (St.-Petersbourg), 83, 5, 125-130. (In Russian).

Probatova, N. S. and Seledets, V. P. (1980). *Rudbeckia hirta* (*Asteraceae*) in Reineke Island (the Primorsky Territory). *Bot. Journ.,* (Leningrad), 65, 7, 977-982. (In Russian).

Probatova, N. S. and Seledets, V. P. (1983). Herbaceous plant communities on Bol'shoi Pelis Island as resources for landscape designing. In: *Landscape designing* (Some problems of theory and methods). Vladivostok: 56-63. (In Russian).

Probatova, N. S. and Seledets, V. P. (1996). Caryological aspect in monitoring of the flora of aquatic and riverside plant communities in the Russian Far East. Abstr. of the Republican conference *"Regional nature management and ecological monitoring"* (276-277). Barnaul: Publishing House of the Altai State University. (In Russian).

Probatova, N.S. and Seledets, V.P. (1997). Problems in coastal plant biodiversity studies and conservation on the Russian Far East. In: V. L. Kassyanov, (Ed.), *Global change studies at the Far East.* Abstr. of Workshop. (29-30). Vladivostok: Dalnauka.

Probatova, N. S. and Seledets, V. P. (1998). Vascular plants in the contact land-ocean zone: the problems of coastal botany in the Russian Far East. In: O. V. Khrapko, (Ed.), *Plants in monsoon climate.* Proceedings of the International conference devoted to the 50-th anniversary of the Botanical Garden-Institute FEB RAS (51-54). Vladivostok: Dalnauka. (In Russian).

Probatova, N. S. and Seledets, V. P. (1999). Vascular plants in the contact "land-ocean" zone. *Herald of the Far East Branch of the Russian Academy of Sciences* (Vladivostok), 3, 80-92. (In Russian).

Probatova, N. S., Seledets, V. P. and Barkalov, V. Yu. (2003). Vascular plants in the contact "land-ocean" zone [the Russian Far East]. First approach // *Phytogeography of Northeast Asia: tasks for the 21-st century.* Abstr. of the Symposium (74). Vladivostok.

Probatova, N. S., Seledets, V. P., Barkalov, V. Yu. and Rudyka, E. G. (2005). Principal results and aspects of the studies on vascular plant biodiversity in the contact land-ocean zone (the Russian Far East). In: O. V. Khrapko, (Ed.), *Rhythms and catastrophes impact on the vegetation of the Russian Far East.* Proceedings of the International conference (112-121). Vladivostok: BGI FEB RAS. (In Russian).

Probatova, N. S., Seledets, V. P., Nedoluzhko, V. A. and Pavlova, N. S. (1998). *Vascular plants in the islands of Peter the Great Bay, Sea of Japan* (the Primorsky Territory). Vladivostok: Dalnauka. (In Russian).

Probatova, N. S., Seledets, V. P. and Sokolovskaya, A. P. (1984). Sea coastal halophilous plants of the soviet Far East: chromosome numbers and ecology. *Komarov Memorial Lectures* (Vladivostok), Issue 31, 89-116. (In Russian).

Probatova, N. S. and Shatokhina, A. V. (2007). Chromosome studies on the Grass Family (Poaceae) in the Amur River basin. In: *Ecosystem evolution of Baikalian region and adjacent areas in context of global change: past, present, future.* Abstr. of International Symposium (71-73). Irkutsk: The Irkutsk State University Publishing House.

Probatova, N. S. and Sokolovskaya, A. P. (1981a). Caryological study on vascular plants from the islands of the Far East state marine reserve. In: Yu. D. Chugunov, (Ed.), *Flowering plants of the islands of the Far East marine reserve* (92-114). Vladivostok: Far East Scientific Center of the USSR Academy of Sciences. (In Russian).

Probatova, N. S. and Sokolovskaya, A. P. (1981b). Chromosome numbers of some species of the Priamurye aquatic and riverside flora, with special reference to peculiarities of its formation. *Bot. Journ.,* (Leningrad), 66, 11, 1584 -1594. (In Russian).

Probatova, N. S. and Sokolovskaya, A. P. (1982). Synopsis of chromosome numbers in *Poaceae* of the Soviet Far East. 1. The tribes *Oryzeae, Brachypodieae, Triticeae. Bot. Journ.,* (Leningrad), 67, 1, 62-70. (In Russian).

Probatova, N. S. and Sokolovskaya, A. P. (1983). New chromosome number data for vascular plants from the islands of Peter the Great Bay (the Primorsky Territory). *Bot. Journ.,* (Leningrad), 68, 12, 1655-1662. (In Russian).

Probatova, N. S. and Sokolovskaya, A. P. (1984). Chromosome numbers in representatives of the families *Alismataceae, Hydrocharitaceae, Hypericaceae, Juncaginaceae, Poaceae, Potamogetonaceae, Ruppiaceae, Sparganiaceae, Zannichelliaceae, Zosteraceae* from the Far East of the USSR. *Bot. Journ.,* (Leningrad), 69, 12, 1700-1702. (In Russian).

Probatova, N. S. and Sokolovskaya, A. P. (1988). Chromosome numbers in vascular plants from the Primorsky Territory, Priamurye, North Koryakia, Kamchatka and Sakhalin. *Bot. Journ.,* (Leningrad), 73, 2, 290-293. (In Russian).

Probatova, N. S. and Sokolovskaya, A. P. (1989). Chromosome numbers in vascular plants from Primorye, Priamurye, Sakhalin, Kamchatka and the Kurile Islands. *Bot. Journ.,* (Leningrad), 74, 1, 120-123. (In Russian).

Probatova, N. S., Sokolovskaya, A. P. and Rudyka, E. G. (1986). Chromosome numbers and distribution of some invasive species and weeds in the Primorsky Territory and Sakhalin. *Proceedings of the Siberian Branch of the USSR Academy of Sciences.* Biol., 13, 2, 63-68. (In Russian).

Probatova, N. S., Sokolovskaya, A. P. and Rudyka, E. G. (1989). Chromosome numbers in some species of vascular plants from Kunashir Isl., the Kurile Islands. *Bot. Journ.,* (Leningrad), 74, 11, 1675-1678. (In Russian).

Probatova, N. S., Sokolovskaya, A. P. and Rudyka, E. G. (1996). Chromosome numbers in species of *Hierochloë* (*Poaceae*) on the Far East of Russia. *Bot. Journ.,* (St.-Petersbourg), 81, 4, 119-121. (In Russian).

Prokopenko S. V. (2000). On the new findings of *Dimeria neglecta* Tzvel. in the Primorsky Territory. In: *Monsoon Climate Plants,* II International conference "Monsoon Climate Plants". (Abstr.). O. V. Khrapko (Ed.). Vladivostok: Dalnauka, (169). (In Russian).

Prokopenko S. V. (2001). New finding of *Ephedra monosperma* in South Primorye. In: Study and designing of lanscapes in the Russian Far East and Siberia, (Collected papers), Vladivostok, Issue 5, (144-149). (In Russian).

Pshenichnikov, B. F. (2003). Individuality of the islands burozems formation and evolution under condition of the monsoon climate of the southern part of the Far East. In: S. B. Gontcharova, (Ed.), *Plants in monsoon climate,* 2003, III. Proceedings of III International conference " Plants in Monsoon Climate" (124-129). Vladivostok: BGI FEB RAS (In Russian).

Pshennikova, L. M. and Berestenko, E. N. (2006). On findings of rare aquatic plants in the Far East. *Bot. Journ.* (St.-Petersbourg), 91, 12, 1919-1921. (In Russian).

Rabotnov, T. A. (1978). Phytocoenology. Moscow University Press. (In Russian).

Rabotnov, T. A. (1995). On the ecological niche of plants. *Ecology,* 3, 246-247. (In Russian).

Rakova, M, V. (1990). *Liparis krameri (Orchidaceae)* - the species new for the flora of USSR from the "Kedrovaya Padj" nature reserve (the Primorsky Territory). *Bot. Journ.* (Leningrad), 75, 12, 1780-1782. (In Russian).

Ramensky, L. G. (1910). On comparative method of ecological study of plant communities. In: *Diary of XII Congress of Russian naturalists and physicians,* 7, 389-390. St.-Petersbourg. (In Russian).

Ramensky, L. G. (1938). *Introduction into complex soil-geobotanical study of lands.* Moscow: Selkhozgiz. (In Russian).

Ramensky, L. G. (1971). *Selected Works. Problems and methods of vegetation study.* Leningrad: Nauka. (In Russian).

Ramensky, L. G. and Tsatsenkin, I. A. (1968). *Ecological evaluation of the forage grasslands in the Caucasus, by vegetation cover.* Moscow: Publishing House of the V.R. Williams All-Union Forage Institute. (In Russian).

Ramensky, L. G., Tsatsenkin, I. A., Chizhikov, O. N. and Antipin, N. A. (1956). *Ecological estimation of grasslands on the base of the vegetation cover study.* Moscow: Selkhozgiz. (In Russian).

Reymers N. F. (1994). *Ecology (theories, laws, regularities, principles and hypotheses).* Moscow: Rossiya Molodaya. (In Russian).

Rodman, L. S., Golub V. P., and Goryaninova, I. N. (1972). The experience of application of L. G. Ramensky scales for evaluation of the meadow vegetation dynamics in southern part of the Volga-Akhtuba flood-lands under conditions of regulated stream // *Transections of the Tymiryazev Agriculture Academy,* 187, 185-191. (In Russian).

Roginsky, A. V. (1988). On the finding of *Atragene coreana* in the south of the Primorsky Territory. *Bulletin of the Main Botanical Garden of Academy of Sciences of the USSR,* Issue 149, (41-43). (In Russian).

Rothera, S. L. and Davy, A. J. (1986). Polyploidy and habitat differentiation in *Deschampsia cespitosa. New Phytol.* 102, 449-467.

Safronova, I. N. and Yurkovskaya, T. K. (2007). Vegetation cover of Pacific Northern Eurasia on geobotanical map scale 1: 15,000,000 for the National Atlas of Russia. In: S. B. Gontcharova, (Ed.), *Plants in monsoon climate,* 2006, IV. Proceedings of IV International conference "Plants in Monsoon Climate" (3-5). Vladivostok: BGI FEB RAS. (In Russian).

Samoilov, Yu. I. (1973). Some results of comparison of ecological scales of Ramensky, Ellenberg, Hundt, and Klapp. *Bot. Journ.* (Leningrad), 58, 5, 646-655. (In Russian).

Sannikova, T. I. (1972). The experience of ecological classification of the flood-land meadows of the Seym River. *Problems in botany:* Transactions of the Kursk Pedagogical Institute, 10 (89), 15-29. (In Russian).

Sannikova, T. I., Paderevskaya, M. I., Kuznetsova, E. A., Makarenko, L. S., Zakharova, V. I., Buyankova, R. I. (1972). Application of ecological scales of the All-Union Forage Institute in Kursk Region. *Problems in botany:* Transactions of the Kursk Pedagogical Institute, 10 (89), 186-188. (In Russian).

Schlothgauer, S. D. (2009). Zoning development of ecosystems regeneration capacities on the example of the Russian Far East central part. In: A. V. Belikovich, (Ed.), *Plants in the Monsoon Climate*, 2009. V. Proceedings of V Conference "Plants in Monsoon Climate" (215-218). Vladivostok: Dalnauka.

Seledets, V. P. (1969). Phytogeographical regionalization of Iturup Island (South Kuriles). In: L. N. Vassilyeva, (Ed.), Problems in botany in the Far East. The Centenary of Academician V. L. Komarov (1869-1969) (181-192). Vladivostok: Far East Branch of Siberian division of the USSR Academy of Sciences. (In Russian).

Seledets, V. P. (1970). On ecology and phytocoenology of coastal zone vegetation in Iturup Island. Proceedings of the Siberian Branch of the USSR Academy of Sciences, *Biol.*, 5, 9-14. (In Russian).

Seledets, V. P. (1975). *Application of ecological scales for study of anthropogenic dynamics of vegetation in suburban areas of the Far East.* In: XII International Botanical Congress. Abstr. of papers, 2 (553). Leningrad. (In Russian).

Seledets, V. P. (1976a). Application of the ecological scales method in botanical studies in the soviet Far East. *Komarov Memorial Lectures* (Vladivostok), 24, 62-78. (In Russian).

Seledets, V. P. (1976b). Ecology of sea coastal grasses [*Poaceae*] in the Far East. Ecology, 2, 19-23. (In Russian).

Seledets, V. P. (1977a). Ecology of the sea coastal plant communities in the Far East. In: *Wild flora of the Far East* (18-33). Vladivostok: Far East Scientific Center of the USSR Academy of Sciences. (In Russian).

Seledets, V. P. (1977b). Recreational successions of the herb layer in *Picea holophylla* forests in the South Primorye. In: *The native flora of the Far East*. Vladivostok: Far East Scientific Center of the USSR Acad. Sci., 1977. P. 62-80. (In Russian).

Seledets, V. P. (1978a). Sea coastal vegetation dynamics in the south of the soviet Far East. In: S. S. Kharkevich (Ed.), *Botanical studies in the Far East*. Transactions of the Institute of Biology and Soil Science, new ser.,

51 (154) (110-115). Vladivostok: Far East Scientific Center of the USSR Academy of Sciences. (In Russian).

Seledets, V. P. (1978b). Anthropogenic dynamics of the herb layer of oak forests in Vladivostok forest park. In: Contemporary problems of nature protection on the Far East (38-43). Vladivostok: Far East Scientific Center of the USSR Academy of Sciences. (In Russian).

Seledets, V.P. 1978c. Application of ecological scales method in botanical studies in the soviet Far East. In: *Komarov Memorial Lectures* (Vladivostok), Issue 24, 38-43. (In Russian).

Seledets, V. P. (1980). Ecological tables for herbaceous plants of Primorye and Priamurye, for phytomelioration. In: *Rational use and protection of land resources of the Far East* (160-170). Vladivostok: Far East Scientific Center of the USSR Academy of Sciences. (In Russian).

Seledets, V. P. (1981). Vegetation of the Bol'shoi Pelis Island. In: Yu. D. Chugunov, (Ed.), *Flowering plants of the islands of the Far East marine reserve* (115-129). Vladivostok: Far East Scientific Center of the USSR Academy of Sciences. (In Russian).

Seledets, V. P. (1982). Application of ecological scales for study of hayfields and pastures of the soviet Far East and measures for their improving. In: S. S. Kharkevich, (Ed.), Wild forage grasses of the soviet Far East (202-229). Moscow: Nauka. (In Russian).

Seledets, V. P. (1985). Ecological and geographical classification of habitats of rare vascular plant species of the soviet Far East. In: *Protection of rare species of vascular plants at the soviet Far East* (181-195). Vladivostok: Far East Scientific Center of the USSR Academy of Sciences. (In Russian).

Seledets, V. P. (1988). Plant communities on the seacoasts of the Far East. In: L.V. Sozinov and B. I. Semkin (Eds.), *Structural organization of ecosystems components* (comparative and quantitative analysis) (35-46). Vladivostok: Far East Branch of the USSR Academy of Sciences. (In Russian).

Seledets, V. P. (2000a). *Ecological scales for vegetation studies in the Russian Far East.* Vladivostok: Publishing House of the Far East Academy of Economics and Management. (In Russian).

Seledets, V.P. (2000b). *Plant cover of the nature monuments in the islands of Peter the Great Bay* (the Primorsky Territory). North Pacific islands biological research (Vladivostok), 4, 1-72 (In Russian).

Seledets, V. P. (2000c). *Anthropogenic dynamics of vegetation of the Russian Far East.* Vladivostok: Pacific Institute of Geography FEB RAS. (In Russian).

Seledets, V. P. (2000d). *The method of ecological scales in botanical studies in the Russian Far East.* Vladivostok: Far East State Academy of Economics and Management (In Russian).

Seledets, V. P. (2000e). Ecological evaluation of habitats and protection of the plant cover in the Far East of Russia. In: O. V. Khrapko, (Ed.), *Monsoon Climate Plants*, 2000. II Intern. conference "Monsoon Climate Plants". (Abstr.), (187-188). Vladivostok: Dalnauka. (In Russian).

Seledets, V. P. (2000f). Experience of application of the ecological scales method to the study of plant cover in monsoon climate of the Russian Far East. In: *Problems in study of the plant cover of Siberia.* 2-d Russian scientific conference devoted to the 150-th anniversary of P. N. Krylov (130-131). Abstr. Tomsk: Tomsk University Press. (In Russian).

Seledets, V. P. (2001). Characteristics and structure of ecological areas of vascular plant species of the Far East, in connection with the problem of biodiversity conservation. In: V. P. Seledets, (Ed.), *V Far-Eastern Conference on nature conservation problems,* devoted to 80-th anniversary of Academician A. V. Zhirmunsky (250-251). Vladivostok: Dalnauka. (In Russian).

Seledets, V. P. (2003a). Ecological differentiation of coastal coenopopulations in the Russian Far East. In: S. B. Gontcharova, (Ed.), *Plants in monsoon climate,* 2003, III. Proceedings of III International conference "Plants in Monsoon Climate" (292-295). Vladivostok: BGI FEB RAS. (In Russian).

Seledets, V. P. (2003b). Ecological ranges of coenotic populations in different parts of the species areas of distribution. In: Phytogeography of Northeast Asia: tasks for the 21-st century. Abstr. of the Symposium (86). Vladivostok.

Seledets, V. P. (2004a). Ecological ranges of species in coastal and continental areas. In: Problems of conservation of the Inner Asia vegetation, I. Proceedings of the scientific conference (10-11). Ulan-Ude: Buryat Scientific Center SB RAS Publishers. (In Russian).

Seledets,V. P. (2004b). Coenopopulation diversity in the nature monuments on the Sea of Japan coast (the Primorsky Territory). *Proceedings of XII-th Meeting of geographers of Siberia and the Russian Far East.* Vladivostok: Pacific Institute of Geography, FEB RAS. (354-356). (In Russian).

Seledets, V. P. (2005a). *Vegetation in the nature monuments on the Sea of Japan coast* (the Primorsky Territory). Vladivostok: Dalnauka.

Seledets, V. P. (2005b). Studies of ecological ranges of plant species in the "continent-ocean" transitional zone: possibilities of the concept of ecological ranges and some results of its application. *The results of the*

Sikhote-Alin nature complexes protection and study: Proceedings of International scientitic and practical conference dedicated to 70-th anniversary of the Sikhote-Alinsky state reserve (Terney, the Primorsky Territory, Sept. 20-23, 2005), Vladivostok (301-308). (In Russian).

Seledets, V. P. (2006). Ecological ranges of plant species at the Russian Pacific coast in comparison with intracontinental regions. *Komarov Memorial Lectures* (Vladivostok), 53, 54-100. (In Russian).

Seledets, V. P. (2008). *Vegetation in the nature monuments of the basin of Peter the Great Bay* (south-west part of the Primorsky Territory), Vladivostok: Botanical Garden-Institute FEB RAS. (In Russian).

Seledets, V. P. (2009a). Comparative ecological study of continental and insular coenopopulations of *Leymus mollis* (Poaceae) in the Russian Far East. *Bot. Journ.,* (St.-Petersbourg), 94, 3, 397-405. (In Russian).

Seledets, V. P. (2009b). Ecological ranges of plant species and their transformation in the contact land-ocean zone (Russian Far East). In: *Ecology of biosystems: the problems of study, indication and forecasting.* Proceedings of II International scientific and practical conference (273-277). Astrakhan: Astrakhansky State University Press. (In Russian).

Seledets, V. P. (2009c). Ecological ranges of the plant species in the continental and ocean regions of Asian Russia. *Sib. Ecol. Journ.* 16, 4, 529-538. (In Russian); *Contemporary Problems of Ecology*, 2, 4, 314-320.

Seledets, V. P. (2010a). Ecological ranges of invasive grass species (Poaceae) in the Russian Far East. *Bot. Journ.,* (St.-Petersbourg), 95, 4, 548-562. (In Russian).

Seledets, V. P. (2010b). Comparative analysis of ecological ranges of *Agrostis trinii* and *Calamagrostis brachytricha* (*Poaceae*) on the east limit of the species geographical distribution (Russian Far East). *Bot. Journ.* (St.-Petersbourg), 95, 9, 1308-1321. (In Russian).

Seledets, V. P. and Probatova, N. S. (1981). Supplement to the flora of the islands of the Far East state marine reserve. In: Yu. D. Chugunov, (Ed.), *Flowering plants of the islands of the Far East marine reserve* (81-91). Vladivostok: Far East Scientific Center of the USSR Academy of Sciences. (In Russian).

Seledets, V. P. and Probatova, N. S. (1989). Polyploidy and ecological-coenotic relationships in plants (on the example of the Soviet Far East flora). In: *2-d Conference on plant caryology.* Abstr. of papers (23-25). Novossibirsk: Central Siberian Botanical Garden, Siberian Division of the USSR Academy of Sciences. (In Russian).

Seledets, V. P. and Probatova, N. S. (1991). Ecological aspects of evolution of *Poaceae* on the soviet Far East. In: *Systematics and evolution in the Grass Family*. Krasnodar: the Kubansky State University, (109-110). (In Russian).

Seledets, V. P. and Probatova, N. S. (2001). Ecological differentiation of *Poaceae* in the Russian Far East. In: *Evolution, Genetics, Ecology and Biodiversity*. International Conference, Symposium "Evolutionary Ideas in Biology", devoted to the memory of Professor Nikolai Vorontsov, (95), Vladivostok: Institute of Biology and Soil Science, Institute of Marine Biology, Vladivostok Public Foundation for Development of Genetics.

Seledets, V. P. and Probatova, N. S. (2003a). Ecological scales as a source of information on ecology of plant diversity (*Poaceae* of the Russian Far East). *Komarov Memorial Lectures* (Vladivostok), 49, 89-130. (In Russian).

Seledets, V. P. and Probatova, N. S. (2003b). Ecological range and some problems of differentiation in the Grass Family (*Poaceae*) of the Russian Far East. In: A. P. Kryukov and L. V. Yakimenko, (Eds.), Problems of Evolution, 5 (213-220). Vladivostok: Dalnauka.

Seledets, V. P. and Probatova, N. S. (2007a). Ecological ranges of plant species: essay of study. In: S. B. Gontcharova, (Ed.), *Plants in monsoon climate*, 2006. IV: Proceedings of IV International conference "Plants in Monsoon Climate" (85-88). Vladivostok: BGI FEB RAS. (In Russian).

Seledets, V. P. and Probatova, N. S. (2007b). *Ecological range of plant species*. Vladivostok: Dalnauka. (In Russian).

Shmida, A. and Elener, S. (1984). Coexistence of plant communities with similar niches. *Vegetatio*, 58, 29-55.

Silvertown, J. W. (1982). *Introduction to plant population ecology*. London, New York: Longman Publ.

Smirnov, A. A. (2009). A new finding of *Cakile edentula* (Bigel.) Hook. (*Brassicaceae*) in the Russian Far East - from the south of Sakhalin. *Bulletin of the Moscow society of nature explorers*, Biol., 114, 6, 72-73. (In Russian).

Smirnova, O. V., Zaugolnova, L. B., and Popadyuk, R. V. (1993). Population concept in biogeocoenology. *Journ. Fundam. Biol.,* 54, 3, 438-448. (In Russian).

Sobolev, L. N. (1971). Delimitation of elementary typological units of plant cover by L. G. Ramensky method. In: *Methods of delimitation of plant associations* (105-110). Leningrad: Nauka. (In Russian).

Sobolev, L. N. (1975). Ecology and typology of the lands. *Ecology*, 4, 20-29. (In Russian).

Sobolev, L. N. (1978). *Method of ecological and typological studies of lands*. Frunze: Ilim. (In Russian).

Sobolev, L. N. and Utekhin, V. D. (1973). Russian (Ramensky) approach to community systematization. In: Handbook of vegetation science, 5, Ordination and classification, (77-102). Hague: Dr. Junk.

Sokolovskaya, A. P. and Probatova, N. S. (1968). Caryotaxonomic study on the Far East species of *Poa* L. *Bot. Journ.*, (Leningrad), 53, 12, 1737-1743. (In Russian).

Sokolovskaya, A. P. and Probatova, N. S. (1973a). Caryotaxonomic study on the Far East species of *Poa* L., II. *Bot. Journ.*, (Leningrad), 58, 1, 89-96. (In Russian).

Sokolovskaya, A. P. and Probatova, N. S. (1973b). Chromosome numbers in the Far East species of *Glyceria* R. Br. *Bot. Journ.*, (Leningrad), 58, 9, 1342-1347. (In Russian).

Sokolovskaya, A. P. and Probatova, N. S. (1974). Caryotaxonomic study on the Far East species of *Agrostis* L. *Bot. Journ.*, (Leningrad), 59, 9, 1278-1287. (In Russian).

Sokolovskaya, A. P. and Probatova, N. S. (1976). Chromosome numbers in the Grass Family of Sakhalin and the Kurile Islands. *Bot. Journ.*, (Leningrad), 61, 3, 384-393. (In Russian).

Sokolovskaya, A. P. and Probatova, N. S. (1977). Caryological study on *Poaceae* in the south part of the soviet Far East. *Bot. Journ.*, (Leningrad), 62, 8, 1143-1153. (In Russian).

Sokolovskaya, A. P. and Probatova, N. S. (1985). Chromosome numbers in vascular plants from the Primorsky Territory, Kamchatskaya Region, the Amur River basin and Sakhalin. *Bot. Journ.*, (Leningrad), 70, 7, 997-999. (In Russian).

Sokolovskaya, A. P. and Probatova, N. S. (1986). Chromosome numbers and distribution of the anthropophilous species, native to Primorsky Territory and the Amur River basin. *Herald of the Leningrad State Univ.*, 2, Biol., 4, 57-63. (In Russian).

Sokolovskaya, A. P., Probatova, N. S. and Rudyka, E. G. (1985). Chromosome numbers in the species of *Asteraceae, Poaceae, Rosaceae* from the Primorsky Territory, Kamchatka and Sakhalin. *Bot. Journ.*, (Leningrad), 70, 1, 126-128. (In Russian).

Sokolovskaya, A. P., Probatova, N. S. and Rudyka, E. G. (1986). The study of the chromosome numbers and distribution of some species of *Lamiaceae* from the Far East of the USSR. *Bot. Journ.*, (Leningrad), 71, 2, 195-200. (In Russian).

Solomon M. E. (1969). *Population dynamics*. The Institute of Biology's Studies in Biology, 18. London.

Stebbins, G. L. (1984). Polyploidy and the distribution of the arctic-alpine flora: new evidence and a new approach. *Bot. Helv.* 94, 1-13.

Stepanova, K. D. (1985). *Meadows of the Kamchatskaya Region.* Vladivostok: Far East Scientific Center of the USSR Academy of Sciences. (In Russian).

Syomkin, B. I., Probatova, N. S., Varchenko, L. I., Gulariants, G. M. (1991). Comparative quantitative analysis of the floristic regions of the soviet Far East, on the base of the Family *Poaceae.* In: *Systematics and evolution in the Grass Family.* Krasnodar: the Kubansky State University, (115-116). (In Russian).

Syomkin, B. I., Seledets, V. P., Mayorov, I. S., Varchenko, L. I., Borzova, L. M. and Gorshkov, V. V. (2010). Methods of comparative study of the components of biodiversity on the botanical nature monuments. *Bot. Journ.*, 95, 3, 408-421. (In Russian).

Takahashi, H. (2009). Geographical distribution patterns of the Apiaceae in Sakhalin and the Kuril Islands. In: H. Takahashi and M. Ŏhara, (Eds.), *Biodiversity and Biogeography of the Kuril Islands and Sakhalin,* 3. (Bulletin of the Hokkaido University Museum, 5) (1-34). Hokkaido University Museum, Sapporo, Japan.

Tateoka, T. (1973). A taxonomic study of the *Poa macrocalyx* complex, with particular reference to the population in Eastern Hokkaido. *Bot. Mag. Tokyo*, 86, 1003, 213-228.

The population structure of vegetation. (1985). Handbook of vegetation science, 3. Dotrecht, Boston, Lankaster.

Tsatsenkin, I. A. (1967). *Ecological scales for grasslands and pastures in mountain and plain regions of Central Asia, Altai and Urals.* Dushanbe: Donish. (In Russian).

Tsatsenkin, I. A. (1970). Ecological evaluation of the forage grasslands in *Carpatian and Balkan Mts. by vegetation cover.* Moscow: V.R. Williams All-Union Research Forage Institute. (In Russian).

Tsatsenkin, I. A., Dmitrieva, S. I., Belyaeva, N. V. and Savchenko, I. V. (1974). *Methodical guideline for ecological evaluation of the forage grasslands in steppe and steppe-forest zones of Siberia by vegetation cover.* Moscow: V.R. Williams All-Union Research Forage Institute. (In Russian).

Tsatsenkin, I. A. and Kassach, A. I. (1970). *Ecological evaluation of grasslands and pastures in Pamirs by vegetation cover.* Dushanbe: Donish. (In Russian).

Tsatsenkin, I. A., Savchenko, I. V. and Dmitrieva, S. I. (1978). *Methodical guideline for ecological evaluation of the forage grasslands in tundra and forest zones of Siberia and the Far East by vegetation cover.* Moscow: V. R. Williams All-Union Research Forage Institute. (In Russian).

Tsyganov, D. N. (1974). Ecomorphes and ecological suites. Bull. Moscow Society of Naturalists, *Biol.* 79, 2, 128-141. (In Russian).

Tsyganov, D. N. (1976). Ecomorphes of the flora of coniferous-deciduous forests. Moscow: *Nauka.* (In Russian).

Tsyganov, D. N. (1983). Phytoindication of ecological regimes in the subzone of mixed forests. Moscow: Nauka. (In Russian).

Tzvelyov, N. N. (1976). Poaceae of the USSR. Leningrad, *Nauka.* (In Russian).

Tzvelyov, N. N. (1983). *Sagittaria aginashi* Makino (*Alismataceae*) - new species for the USSR. *Bot. Journ.* (Leningrad), 68, 11, 1469-1470. (In Russian).

Tzvelyov, N. N. (1985a). A species new for the flora of USSR - Eleocharis tetraquetra Nees (*Cyperaceae*) from the Far East. Novitates systematicae plantarum vascularium, 22, 44-46. Leningrad: *Nauka.* (In Russian).

Tzvelyov, N. N. (1985b). Three new species of the genus *Eriocaulon* (*Eriocaulaceae*) from the Far East. *Bot. Journ.* (Leningrad), 70, 3, 390-394. (In Russian).

Tzvelyov, N. N. (2008). On the genus *Elymus* L. (*Poaceae*) in Russia. *Bot. Journ.* (St.-Petersbourg), 93, 10, 1587-1596. (In Russian).

Tzvelyov, N. N. (2009). Critical notes on *Poaceae* in Russia. *Bot. Journ.* (St.-Petersbourg), 94, 2, 275-282. (In Russian).

Tzvelyov, N. N. and Probatova, N. S. (2009). New data in *Poaceae* of the flora of the Russian Far East. In: A. V. Belikovich, (Ed.), *Plants in Monsoon Climate V*: Proceedings of V Conference "Plants in Monsoon Climate" (179-183). Vladivostok: Dalnauka. (In Russian).

Tzvelyov, N. N. and Probatova, N. S. (2010a). The genera *Elymus* L., *Elytrigia* Desv., *Agropyron* Gaertn., *Psathyrostachys* Nevski and *Leymus* Hochst. (*Poaceae: Triticeae*) in the flora of Russia. *Komarov Memorial Lectures* (Vladivostok), 57, 5-102. (In Russian).

Tzvelyov, N. N. and Probatova, N. S. (2010b). New taxa of *Poaceae* from Russia. *Bot. Zhurn.,* (St.-Petersbourg), 95, 6, 131-141. (In Russian).

Utyasheva, T. R. (2005). On the state of the *Eremurus altaicus* (Pall.) Stev. (*Asphodeliaceae*) populations in Markakol depression. In: *Biodiversity and spatial organization of vegetation in Siberia, methods of study and protection*: Proceedings of All-Russian conference (Novossibirsk, 25-27 Oct. 2005). Novossibirsk, Central Siberian Botanical Garden, Siberian Branch of the Russian Academy of Sciences. (146-148). (In Russian).

Vascular plants of the Soviet Far East. (1985-1996). S. S. Kharkevich, (Ed.).Vol. 1, 1985; vol. 2, 1987; vol. 3, 1988; vol. 4, 1989; vol. 5, 1991;

vol. 6, 1992; vol. 7, 1995; vol. 8, 1996. Leningrad-St.-Petersbourg: Nauka. (In Russian).

Whittaker R. H. (1975). *Communities and ecosystems.* 2 ed. New-York.

Woroshilov, V. N. (1968). On the riverbank flora of temperate regions in monsoon climate. *Bulletin of the Main Botanical Garden of Academy of Sciences of the USSR*, Issue 68 (45-48). (In Russian).

Yaroshenko, P. D. (1955). On the similarity of the *Arundinella* grasslands on Khanka Plain with some types of prairies. In: Transactions of the Far East Branch of the USSR Academy of Sciences, Issue 8, (41-44). (In Russian).

Yaroshenko, P. D. (1958). Forest - steppe of the soviet Far East and ajacent areas of North-East China. In: Problems of agriculture and forestry in the Far East, Issue 2. (203-219). (In Russian).

Yurtsev, B. A. (1987). Plant populations as a matter of geobotany, floristics and phytogeography. *Bot. Journ.* (Leningrad), 72, 5, 581-588. (In Russian).

Zaugolnova, L. B. (1976). Types of the age structures of normal plant coenopopulations. In: *Plant coenopopulations.* Moscow: *Nauka*, (81-91). (In Russian).

Zlobin, Yu. A. (1989). *Principles and methods of study of plant coenotic populations.* Kazan: The Kazan Univ. Press. (In Russian).

Zlobin, Yu. A. (1996). The structure of plant populations. *Progress in contemporary biology*, 116, 2, 133-146. (In Russian).

Brachybotrys paridiformis

Chloranthus jopanicus

Fritillaria ussuriensis

Hylomecon vernalis

Lonicera praeflorens

October, near Vladivostok.

Plagiorhegma dubia

Rosa rugosa

Russky Island

Viola orientalis

APPENDIX 1.

Appendix 1. Ecological scales for vascular plant species of East Siberia and the Russian Far East

Scales (total number of grades):
H (humidity, 120), VH (variability of humidity, 20), D (drainage, 12), SFS (soil fertility and salinity, 30), G (granulometric composition of soil, 15), R (renewal of soil, 20), Sh (shading, 15), RPL (recreational and pasture load, 10) . (+) alien in the RFE.

Species	Scale	Abundance (%)				
		>8 massales (m)	2,5-8 copiosae (c)	0,2-2,5 numerosae (n)	0,1-0,2 pauces (p)	<0,1 solitariae (s)
1	2	3	4	5	6	7
Achnatherum extremiorientale (Hara) Keng ex Tzvel. (*Poaceae*)	H SFS RPL	56-68 6-12 1-2	56-69 5-13 1-3	54-71 4-15 1-4	50-75 1-16 1-5	48-76 1-16 1-6
Achnatherum sibiricum (L.) Keng ex Tzvel.	H SFS RPL	-	-	49-60 10-14 2-5	49-60 10-14 2-6	40-61 10-15 2-7

Appendix 1. (Continued)

Species	Scale	Abundance (%)				
		>8 massales (m)	2,5-8 copiosae (c)	0,2-2,5 numerosae (n)	0,1-0,2 pauces (p)	<0,1 solitariae (s)
1	2	3	4	5	6	7
Agrostis anadyrensis Socz. (*Poaceae*)	H	65-72	62-74	60-78	58-82	54-84
	VH	1-4	1-5	1-6	1-7	1-8
	D	4-7	4-7	3-7	3-7	3-7
	SFS	7-8	6-9	5-10	5-10	4-11
	G	6-10	5-11	5-11	5-11	4-12
	R	14-17	13-18	12-19	12-19	11-20
	Sh	3-5	3-6	2-6	2-7	1-8
	RPL	1-3	1-4	1-5	1-7	1-8
+*Agrostis capillaris* L.	H	61-81	61-81	60-82	59-89	58-90
	SFS	8-17	8-17	8-17	8-18	8-18
	RPL	2-3	2-5	2-8	1-9	1-9
Agrostis clavata Trin.	H	58-72	53-75	49-78	45-80	40-85
	VH	7-11	6-12	4-13	3-14	2-15
	D	3-9	2-10	2-10	2-11	1-12
	SFS	8-10	6-12	4-13	3-14	1-16
	G	3-8	2-9	2-10	1-11	1-12
	RPL	1-4	1-5	1-6	1-8	1-10
+*Agrostis gigantea* Roth	H	60-92	60-96	58-99	51-101	51-101
	SFS	10-13	9-14	8-17	5-21	5-21
	RPL	2-4	2-5	1-7	1-9	1-9
Agrostis scabra Willd.	H	62-74	61-75	60-76	59-77	57-78
	VH	4-11	3-12	2-13	2-14	2-15
	D	1-5	1-6	1-6	1-6	1-7
	SFS	7-12	6-13	6-14	5-15	4-16
	G	6-8	5-9	4-10	4-11	3-12
	R	15-18	12-18	10-18	7-18	4-19
	Sh	1-2	1-3	1-3	1-3	1-4
	RPL	1-5	1-6	1-7	1-8	1-8

Appendix 1

Species	Scale	Abundance (%)				
		>8 massales (m)	2,5-8 copiosae (c)	0,2-2,5 numerosae (n)	0,1-0,2 pauces (p)	<0,1 solitariae (s)
1	2	3	4	5	6	7
Agrostis stolonifera L.	H	69-93	67-95	66-100	64-102	62-104
	SFS	11-13	9-15	9-17	9-19	9-19
	RPL	1-3	1-4	1-4	1-6	1-6
Agrostis trinii Turcz.	H	58-80	53-82	51-83	46-87	46-87
	SFS	9-15	8-16	8-16	7-17	7-17
	RPL	1-5	1-5	1-6	1-7	1-7
Alopecurus glaucus Less. (*Poaceae*)	H	72-76	70-78	69-79	68-71	66-82
	VH	6-12	5-13	5-13	4-14	3-15
	D	1-4	1-5	1-5	1-6	1-7
	SFS	7-9	6-10	5-11	4-12	2-14
	G	2-5	3-7	3-9	3-10	3-13
	R	14-17	13-17	12-18	10-19	9-19
	RPL	1-2	1-3	1-5	1-6	1-7
Arctagrostis latifolia (R. Br.) Griseb. (*Poaceae*)	H	66-79	69-83	60-86	57-90	54-92
	VH	4-8	3-10	2-12	1-13	1-16
	D	1-7	1-9	1-10	1-11	1-12
	SFS	5-10	4-11	3-12	3-12	2-13
	G	5-12	5-12	4-13	4-13	4-13
	R	8-12	6-14	5-15	3-17	1-19
	Sh	3-5	3-5	2-5	2-6	1-6
Arctophila fulva (Trin.) Anderss. (*Poaceae*)	H	77-88	70-92	70-94	65-98	62-102
	VH	1-6	1-8	1-9	1-10	1-12
	D	1-2	1-3	1-3	1-4	1-5
	SFS	5-13	4-14	4-14	3-14	2-15
	G	4-9	3-10	2-11	1-12	1-12
	R	14-18	13-19	13-19	12-20	12-20
	Sh	2-3	2-3	2-3	1-4	1-4
	RPL	1-3	1-4	1-4	1-5	1-6
Arctopoa eminens (J. S. Presl) Probat. (*Poaceae*)	H	61-94	60-96	58-97	57-98	56-99
	VH	1-4	1-6	1-6	1-7	1-7
	SFS	4-20	3-21	2-21	2-22	1-22
	G	5-6	5-6	5-6	5-6	4-6
	Sh	1-3	1-3	1-3	1-3	1-3
	RPL	1-2	1-3	1-4	1-5	1-7

Appendix 1. (Continued)

Species	Scale	Abundance (%)				
		>8 massales (m)	2,5-8 copiosae (c)	0,2-2,5 numerosae (n)	0,1-0,2 pauces (p)	<0,1 solitariae (s)
1	2	3	4	5	6	7
Arctopoa subfastigiata (Trin.) Probat.	H	-	-	60-63	57-71	57-71
	SFS	-	-	10-14	10-14	10-14
	RPL	-	-	2-5	2-5	2-5
Artemisia keiskeana Miq. (*Asteraceae*)	H	63-67	61-68	58-72	53-77	53-77
	SFS	8-10	5-13	2-16	1-18	1-18
	RPL	1-2	1-3	1-4	1-6	1-6
Artemisia mandshurica (Kom.) Kom.	H	60-65	57-67	55-71	49-77	49-77
	SFS	9-13	8-14	7-15	5-17	5-17
	RPL	1-3	1-4	1-5	1-6	1-7
Artemisia stolonifera (Maxim.) Kom.	H	63-71	62-72	61-73	59-75	59-75
	SFS	10-14	9-15	9-15	7-17	7-17
	RPL	1-3	1-4	1-5	1-8	1-8
Arundinella anomala Steud. (*Poaceae*)	H	56-65	52-69	52-69	52-77	52-77
	SFS	10-12	10-13	10-13	8-13	8-13
	RPL	2-3	2-3	2-3	2-3	2-3
Arundinella hirta (Thunb.) Tanaka	H	56-68	52-69	52-70	50-77	50-77
	SFS	6-14	6-14	5-15	4-15	4-15
	RPL	1-3	1-4	1-5	1-6	1-6
Avenella flexuosa (L.) Drej. (*Poaceae*)	H	64-70	61-80	61-83	60-85	60-85
	SFS	5-9	5-9	4-10	4-10	4-10
	RPL	2-3	2-3	2-3	2-4	2-4
Avenula dahurica (Kom.) Holub (*Poaceae*)	H	-	-	58-69	56-71	56-71
	SFS			10-11	8-13	8-13
	RPL			1-4	1-6	1-6
Avenula schelliana (Hack.) Holub	H	-	-	51-62	50-64	50-64
	SFS	-	-	6-10	4-12	4-12
	RPL	-	-	2-3	2-3	2-3

Species	Scale	Abundance (%)				
		>8 massales (m)	2,5-8 copiosae (c)	0,2-2,5 numerosae (n)	0,1-0,2 pauces (p)	<0,1 solitariae (s)
1	2	3	4	5	6	7
Beckmannia syzigachne (Steud.) Fern. (*Poaceae*)	H	64-99	62-100	61-100	57-100	57-100
	SFS	8-13	8-15	8-16	6-17	6-17
	RPL	1-4	1-4	1-5	1-9	1-9
Bromopsis canadensis (Michx.) Holub. (*Poaceae*)	H	69-75	68-75	66-80	65-81	63-83
	VH	12-15	11-16	10-17	8-18	7-20
	D	4-6	3-7	3-7	2-8	1-9
	SFS	8-11	7-12	5-14	4-15	2-20
	G	6-7	5-8	4-8	3-8	1-11
	R	14-17	13-20	12-20	12-20	10-20
	Sh	3-5	3-6	2-7	2-7	1-8
	RPL	1-2	1-3	1-4	1-5	1-6
+*Bromopsis inermis* (Leys.) Holub	H	58-73	57-76	51-88	49-88	49-88
	SFS	10-13	9-13	8-13	8-17	8-17
	RPL	3-5	2-7	2-8	1-9	1-9
Bromopsis pumpelliana (Scribn.) Holub	H	64-72	62-75	60-77	57-79	54-84
	D	4-9	3-10	2-11	1-11	1-12
	SFS	8-13	7-15	5-16	4-18	2-20
	G	5-10	4-10	3-12	2-13	1-14
	RPL	1-4	1-5	1-6	1-6	1-7
Bupleurum longiradiatum Turcz. (*Apiaceae*)	H	60-68	58-68	55-74	51-75	51-75
	SFS	6-12	5-13	4-14	2-16	2-16
	RPL	1-2	1-3	1-4	1-4	1-4
Calamagrostis angustifolia Kom. (*Poaceae*)	H	60-89	60-90	60-92	60-92	60-92
	SFS	9-14	8-14	5-15	5-15	3-15
	RPL	2-4	2-5	2-5	2-5	2-5
Calamagrostis brachytricha Steud.	H	58-68	56-68	54-70	50-74	50-74
	SFS	8-12	7-13	7-13	6-14	4-16
	RPL	1-3	1-4	1-4	1-5	1-5
Calamagrostis deschampsioides Trin	H	66-88	64-90	63-91	61-92	59-93
	VH	1-2	1-2	1-3	1-3	1-4
	D	1-3	1-3	1-3	1-4	1-4
	SFS	7-19	6-20	-12	5-22	4-24
	G	4-9	3-10	2-11	2-11	1-12
	R	9-11	7-13	5-15	4-16	2-18
	Sh	3-4	3-4	3-4	2-5	2-5
	RPL	1-2	1-3	1-4	1-5	1-6

Appendix 1. (Continued)

Species	Scale	Abundance (%)				
		>8 massales (m)	2,5-8 copiosae (c)	0,2-2,5 numerosae (n)	0,1-0,2 pauces (p)	<0,1 solitariae (s)
1	2	3	4	5	6	7
Calamagrostis langsdorffii (Link) Trin.	H	59-92	57-94	54-95	52-99	52-99
	SFS	8-18	7-19	5-20	3-23	3-23
	RPL	1-6	1-7	1-8	1-10	1-10
Calamagrostis neglecta (Ehrh.) Gaertn., Mey. et Schreb.	H	65-90	65-98	**53-100**	61-105	61-105
	VH	7-12	7-12	7-12	7-12	7-12
	SFS	4-14	3-14	3-15	3-15	3-15
	R	3-5	3-5	3-5	3-5	3-5
	RPL	1-4	1-4	1-4	1-4	1-4
Calamagrostis purpurascens R.Br.	H	70-76	68-76	67-82	66-84	65-87
	D	10-11	10-11	9-11	8-11	8-12
	SFS	4-6	3-7	3-8	3-8	2-9
	G	8-10	7-11	6-12	5-13	4-14
	RPL	1-4	1-4	1-5	1-5	1-6
Calamagrostis sachalinensis Fr. Schmidt	H	64-75	60-76	56-76	56-76	56-76
	VH	7-8	5-8	5-8	5-8	5-8
	SFS	8-10	8-10	7-10	7-10	7-10
	R	1-5	1-5	1-5	1-5	1-5
	RPL	3-5	3-5	2-6	2-6	2-6
Calamagrostis sesquiflora (Trin.) Tzvel.	H	69-72	66-75	64-78	62-82	58-84
	VH	8-12	7-13	6-14	5-15	4-16
	D	9-12	8-12	7-12	5-12	4-12
	SFS	3-7	2-8	2-9	1-9	1-10
	G	10-12	10-12	9-13	8-13	8-14
	R	1-5	1-6	1-7	1-8	1-9
	Sh	2-5	2-6	2-6	1-7	1-8
	RPL	1-2	1-3	1-4	1-6	1-8

Appendix 1 117

Species	Scale	Abundance (%)				
		≥8 massales (m)	2,5-8 copiosae (c)	0,2-2,5 numerosae (n)	0,1-0,2 pauces (p)	<0,1 solitariae (s)
1	2	3	4	5	6	7
Carex appendiculata (Trautv. et Mey.) Kiik. (*Cyperaceae*)	H	61-93	61-98	61-99	59-100	59-100
	SFS	6-13	6-14	6-14	6-14	6-14
	RPL	2-4	2-4	2-4	2-4	2-4
Carex campylorhina V. Krecz.	H	59-71	57-72	56-73	53-77	53-77
	SFS	6-10	5-11	4-12	2-14	2-14
	RPL	1-3	1-4	1-5	1-6	1-6
Carex falcata Turcz.	H	62-70	62-70	61-72	53-76	53-76
	SFS	9-11	9-11	6-11	4-14	4-14
	RPL	2-3	2-3	2-3	1-5	1-5
Carex nanella Ohwi	H	59-69	57-71	56-72	52-76	52-76
	SFS	6-12	6-12	5-13	4-14	4-14
	RPL	1-3	1-3	1-4	1-9	1-9
Carex pallida C.A. Mey.	H	61-70	57-70	57-74	57-75	57-75
	SFS	8-12	6-12	6-12	6-12	6-12
	RPL	2-3	2-4	2-3	2-4	2-4
Carex schmidtii Meinsh.	H	59-98	59-97	57-97	54-99	54-99
	SFS	5-13	5-13	5-14	3-15	3-15
	RPL	2-7	2-7	2-7	1-8	1-8
Carex siderosticta Hance	H	68-76	68-78	65-79	61-83	61-83
	SFS	6-12	5-13	4-14	3-15	3-15
	RPL	1-5	1-5	1-6	1-9	1-9
Carex ussuriensis Kom.	H	58-68	56-67	55-69	52-72	52-72
	SFS	9-13	8-14	7-15	5-17	5-17
	RPL	3-7	3-7	2-8	1-9	1-9
Chloranthus japonicus Siebold. (*Chloranthaceae*)	H	70-76	69-77	67-79	64-82	64-82
	SFS	7-13	7-13	6-14	4-16	4-16
	RPL	1-2	1-2	1-3	1-4	1-4
Cimicifuga simplex (Wormsk. ex DC) Turcz. (*Ranunculaceae*)	H	-	64-68	63-72	60-78	60-78
	SFS		8-10	8-10	8-12	8-12
	RPL		2-3	1-3	1-4	1-4
Cleistogenes kitagawae Honda (*Poaceae*)	H	50-56	48-58	45-61	40-63	40-63
	SFS	9-15	8-16	7-17	5-19	5-19
	RPL	1-3	1-4	1-5	1-9	1-9

Appendix 1. (Continued)

Species	Scale	Abundance (%)				
		>8 massales (m)	2,5-8 copiosae (c)	0,2-2,5 numerosae (n)	0,1-0,2 pauces (p)	<0,1 solitariae (s)
1	2	3	4	5	6	7
+*Dactylis glomerata* L. (*Poaceae*)	H	60-70	60-72	60-74	59-78	59-78
	SFS	8-12	8-13	8-14	-15	7-15
	RPL	2-3	2-3	2-3	2-3	2-3
Danthonia riabuschinskii (Kom.) Kom. (*Poaceae*)	H	57-84	56-84	53-87	50-97	50-97
	SFS	9-10	8-12	6-14	4-17	4-17
	RPL	2-3	2-4	2-6	1-8	1-8
Elymus confusus (Roshev.) Tzvel. (*Poaceae*)	H	67-80	67-80	55-80	50-80	50-80
	VH	10-12	8-12	3-12	1-12	1-12
	SFS	9-13	7-13	4-13	3-13	3-13
	R	1-2	1-2	1-5	1-7	1-7
	RPL	1-5	1-8	1-9	1-10	1-10
Elymus gmelinii (Ledeb.) Tzvel.	H	65-67	65-70	57-70	57-75	55-80
	VH	9-12	8-12	7-12	6-12	5-13
	SFS	9-12	9-12	9-12	9-12	7-12
	RPL	2-6	2-6	2-7	2-7	2-7
Elymus jacutensis (Drob.) Tzvel.	H	60-67	60-68	52-70	50-70	50-70
	VH	2-11	2-11	2-11	2-12	2-12
	SFS	12-13	11-15	6-15	6-16	6-16
	R	1-9	2-9	3-9	4-9	5-9
	RPL	1-6	1-6	2-8	2-8	2-9
Elymus kamczadalorum (Nevski) Tzvel.	H	66-69	66-69	66-70	65-71	65-71
	VH	7-10	7-10	7-11	7-11	7-11
	SFS	7-9	7-10	7-12	7-12	7-13
	R	1-5	1-5	1-6	1-6	1-6
	RPL	3-5	3-5	2-5	2-5	2-5
Elymus kronokensis (Kom.) Tzvel.	H	62-72	62-72	60-72	55-74	45-75
	VH	8-9	8-9	8-10	8-12	8-16
	SFS	9-14	9-15	9-17	8-17	7-19
	RPL	4-7	4-7	4-8	2-8	2-8
Elymus macrourus (Turcz.) Tzvel.	H	-	-	60-72	60-72	60-72
	VH			8-10	8-10	8-10
	SFS			2-4	2-4	2-4
	RPL			1-7	1-7	1-7

Species	Scale	Abundance (%)				
		>8 massales (m)	2,5-8 copiosae (c)	0,2-2,5 numerosae (n)	0,1-0,2 pauces (p)	<0,1 solitariae (s)
1	2	3	4	5	6	7
Elymus sibiricus L.	H	57-72	55-74	55-74	55-83	55-83
	SFS	11-13	10-14	5-14	5-16	5-16
	RPL	2-3	2-3	2-6	2-7	2-7
Elytrigia repens (L.) Nevski (*Poaceae*)	H	50-77	50-83	44-90	42-96	42-96
	SFS	9-16	9-16	9-16	9-19	9-19
	RPL	1-6	1-8	1-9	1-9	1-9
Eupatorium lindleyanum DC. (*Asteraceae*)	H	-	58-70	60-71	60-97	60-97
	SFS		11-12	10-13	10-14	10-14
	RPL		2-3	2-3	2-4	2-4
Festuca altaica Trin. (*Poaceae*)	H	59-70	59-73	59-83	58-73	58-83
	SFS	8-10	6-10	6-11	5-11	5-11
	RPL	1-3	1-3	2-3	2-3	2-4
Festuca brachyphylla Schult. et Schult. fil.	H	-	58-79	56-79	56-79	56-83
	SFS		5-10	5-11	4-12	4-12
	RPL		2-3	2-3	2-4	2-4
Festuca extremiorientalis Ohwi	H	60-67	57-69	56-72	49-77	49-77
	SFS	12-15	11-17	10-18	9-19	9-19
	RPL	1-2	1-3	1-4	1-6	1-6
Festuca ovina L.	H	60-66	55-68	53-77	50-79	50-80
	SFS	8-11	7-13	5-15	5-15	5-15
	RPL	2-4	2-5	2-5	2-6	2-6
Festuca rubra L.	H	56-70	50-75	50-80	50-87	50-87
	SFS	8-14	7-15	7-15	5-19	5-19
	RPL	3-5	3-5	3-6	1-7	1-7
Filipendula camtschatica (Pall.) Maxim. (*Rosaceae*)	H	63-73	63-73	62-77	62-78	62-78
	SFS	8-10	8-10	8-11	8-12	8-12
	RPL	2-3	1-4	1-4	1-4	1-4
Filipendula palmata (Pall.) Maxim.	H	74-84	71-87	69-89	64-94	64-94
	SFS	7-13	6-14	5-15	3-17	3-17
	RPL	1-3	1-4	1-5	1-6	1-6
Fimbripetalum radians (L.) Ikonn. (*Caryophyllaceae*)	H	72-88	70-90	69-92	64-94	64-96
	SFS	10-16	9-17	9-17	7-18	7-18
	RPL	1-3	1-4	1-5	3-8	3-8
Fragaria orientalis Losinsk. (*Rosaceae*)	H	61-71	58-74	55-76	50-82	50-82
	SFS	5-13	4-14	4-14	2-16	2-16
	RPL	2-6	2-6	1-6	1-8	1-8

Appendix 1. (Continued)

Species	Scale	Abundance (%)				
		>8 massales (m)	2,5-8 copiosae (c)	0,2-2,5 numerosae (n)	0,1-0,2 pauces (p)	<0,1 solitariae (s)
1	2	3	4	5	6	7
Galium boreale L. (Rubiaceae)	H	63-71	60-74	58-71	53-81	53-81
	SFS	8-14	8-14	7-15	6-16	6-16
	RPL	1-4	1-5	1-5	1-7	1-7
Galium verum L.	H	57-67	53-72	50-74	43-76	43-76
	SFS	8-14	7-15	7-15	8-17	8-17
	RPL	2-6	1-6	1-6	1-8	1-8
Geranium erianthum DC. (Geraniaceae)	H	62-69	62-73	62-73	60-86	60-86
	SFS	8-11	8-11	8-11	6-11	6-11
	RPL	2-3	2-3	2-3	1-3	1-3
Geranium eriostemon Fisch.	H	67-75	65-77	64-78	60-82	60-82
	SFS	10-14	10-14	9-15	8-16	8-16
	RPL	1-3	1-4	1-5	1-7	1-7
Geranium sibiricum L.	H	55-65	53-68	51-69	51-69	51-69
	SFS	9-13	8-14	7-15	6-16	6-16
	RPL	4-6	3-7	3-7	2-8	2-8
Geranium wlassowianum Fisch. ex Link	H	75-81	73-84	58-87	50-90	50-90
	SFS	10-16	10-16	10-16	8-18	8-18
	RPL	1-3	1-4	1-4	1-6	1-6
Geum aleppicum Jacq. (Rosaceae)	H	65-71	63-78	60-75	58-80	58-80
	SFS	9-15	9-15	8-16	7-17	7-17
	RPL	3-7	3-7	2-8	7-9	7-9
Glyceria lithuanica (Gorski) Gorski (Poaceae)	H	-	-	-	78-79	78-79
	VH				7-9	7-9
	SFS				9-10	9-10
	R				4-5	4-5
	RPL				1-2	1-2
Glyceria spiculosa (Fr. Schmidt) Roshev.	H	97-100	95-102	91-102	91-103	91-103
	SFS	8-13	6-14	6-14	6-15	6-15
	RPL	2-3	1-3	1-3	1-3	1-3
Glyceria triflora (Korsh.) Kom.	H	-	87-96	83-97	58-90	58-90
	SFS		7-11	7-12	5-13	5-13
	RPL		2-4	2-4	2-4	2-4

Appendix 1

Species	Scale	Abundance (%)				
		>8 massales (m)	2,5-8 copiosae (c)	0,2-2,5 numerosae (n)	0,1-0,2 pauces (p)	<0,1 solitariae (s)
1	2	3	4	5	6	7
Glycine soja Siebold et Zucc. (*Fabaceae*)	H	-	61-65	59-72	55-72	55-72
	SFS		10-11	9-12	5-16	5-16
	RPL		1-3	1-5	108	1-8
Gypsophila pacifica Kom. (*Caryophyllaceae*)	H	56-62	55-63	53-65	50-68	50-68
	SFS	8-14	7-15	7-15	5-17	5-17
	RPL	2-6	2-6	2-6	1-7	1-7
Hieracium umbellatum L. (*Asteraceae*)	H	-	56-80	56-80	55-80	55-80
	SFS		10-13	9-13	8-14	8-14
	RPL		2-4	1-4	1-6	1-6
Hierochloë alpina (Sw.) Roem. et Schult. (*Poaceae*)	H	70-72	69-85	69-86	67-87	67-87
	VH	8-9	8-9	8-9	8-9	8-9
	SFS	5-11	4-11	3-11	2-11	1-11
	R	8-9	8-9	8-9	8-9	8-9
	RPL	2-3	2-3	2-3	2-3	2-3
Honckenya oblongifolia Torr. et Gray (*Caryophyllaceae*)	H	80-87	76-88	73-89	69-90	65-91
	VH	4-6	3-6	3-7	3-7	2-8
	D	5-6	5-6	5-6	4-6	4-6
	SFS	18-20	17-20	16-20	14-21	13-21
	G	8-9	7-10	6-11	5-11	4-12
	Sh	1-2	1-2	1-3	1-3	1-4
	RPL	1-3	1-4	1-5	1-6	1-7
Hypericum gebleri Ledeb. (*Hypericaceae*)	H	68-72	64-84	60-86	52-94	52-94
	SFS	7-13	6-14	5-15	4-16	4-16
	RPL	1-2	1-3	1-4	1-5	1-5
Iris ensata Thunb. (*Iridaceae*)	H	76-82	74-84	72-86	67-91	67-91
	SFS	11-15	11-15	11-15	9-17	9-17
	RPL	1-2	1-3	1-4	1-5	1-6
Iris setosa Pall. ex Link	H	62-83	62-83	58-92	58-102	58-102
	SFS	8-13	8-13	6-14	6-14	6-14
	RPL	1-3	2-3	2-4	2-4	2-4
Juncus bufonius L. (*Juncaceae*)	H	65-84	65-86	63-86	61-90	61-90
	SFS	7-10	5-12	2-8	1-10	1-10
	RPL	2-4	2-6	2-8	1-10	1-10
Kalimeris incisa (Fisch.) DC. (*Asteraceae*)	H	60-69	60-69	60-71	60-71	60-71
	SFS	10-12	10-12	10-12	10-12	10-12
	RPL	2-3	2-3	2-3	2-3	2-3

Appendix 1. (Continued)

Species	Scale	Abundance (%)				
		>8 massales (m)	2,5-8 copiosae (c)	0,2-2,5 numerosae (n)	0,1-0,2 pauces (p)	<0,1 solitariae (s)
1	2	3	4	5	6	7
Koeleria cristata (L.) Pers. (*Poaceae*)	H	50-64	50-66	44-77	42-79	42-79
	SFS	11-13	11-14	7-14	5-18	5-18
	RPL	2-5	2-6	2-7	2-8	2-9
Lathyrus japonicus Willd. (*Fabaceae*)	H	64-73	62-74	61-75	59-75	57-76
	VH	5-8	5-9	4-10	3-11	2-12
	D	5-6	5-6	5-6	5-6	5-6
	SFS	14-20	11-20	9-21	6-21	4-21
	G	6-9	6-10	6-11	6-12	6-14
	R	15-19	15-19	15-19	14-19	14-20
	Sh	1-2	1-2	1-3	1-3	1-4
	RPL	1-5	1-5	1-6	1-6	1-7
Lespedeza juncea (L. fil.) Pers. (*Fabaceae*)	H	52-60	52-60	52-61	50-63	50-63
	SFS	11-13	11-13	10-14	10-14	10-14
	RPL	2-4	2-4	2-6	206	2-6
+*Leymus chinensis* (Trin.) Tzvel. (*Poaceae*)	H	53-60	53-61	53-62	53-63	53-64
	SFS	12-13	11-13	10-13	8-15	8-15
	RPL	2-3	2-3	2-4	1-6	1-6
Leymus mollis (Trin.) Hara	H	73-93	65-93	65-94	65-95	65-97
	D	5-6	5-6	5-6	4-6	1-6
	SFS	17-22	15-22	14-22	14-22	13-22
	G	6-9	6-10	6-11	6-11	6-12
	RPL	1-2	1-3	1-4	1-5	1-7
Ligusticum hultenii Fern. (*Apiaceae*)	H	67-70	65-73	61-74	63-75	61-77
	VH	4-6	4-6	3-7	3-8	2-8
	D	5-7	5-9	5-10	4-11	4-12
	SFS	9-11	9-13	9-15	9-16	8-19
	R	15-17	14-18	13-18	12-19	11-20
	G	5-11	5-12	5-12	4-14	4-15
	Sh	2-3	2-4	2-4	1-5	1-6
	RPL	2-3	2-4	2-4	1-5	1-6
Lilium pensylvanicum Ker-Gawl. (*Liliaceae*)	H	-	58-70	58-70	57-73	58-73
	SFS		10-12	10-12	9-13	9-13
	RPL		2-3	2-3	2-4	2-4

Appendix 1

Species	Scale	Abundance (%)				
		>8 massales (m)	2,5-8 copiosae (c)	0,2-2,5 numerosae (n)	0,1-0,2 pauces (p)	<0,1 solitariae (s)
1	2	3	4	5	6	7
+*Linaria vulgaris* Mill. (*Scrophulariaceae*)	H SFS RPL	49-71 9-13 2-5	49-71 9-13 2-5	51-73 11-15 2-7	50-79 10-17 2-8	50-79 10-17 2-8
Lysimachia davurica Ledeb. (*Primulaceae*)	H SFS RPL	73–79 10–14 1–2	69–83 10–14 1–3	66–83 10–14 1–4	58–95 9–15 1–7	58–95 9–15 1–7
Lythrum salicaria L. (*Lythraceae*)	H SFS RPL	75–85 10–16 1–6	72–88 10–16 1–6	69–90 9–17 1–6	64–94 9–19 1–7	64–96 9–19 1–7
Mertensia simplicissima (Ledeb.) G. Don fil. (*Boraginaceae*)	H VH D SFS G R Sh RPL	64–76 1–4 5–6 19–20 6–9 13–16 2–3 1–4	63–77 1–4 5–6 16–20 6–10 13–16 2–4 1–5	61–77 1–4 5–6 12–21 6–11 13–17 2–4 1–5	59–80 1–4 5–6 9–21 5–11 12–19 1–5 1–6	57–82 1–4 4–7 5–22 5–12 11–20 1–6 1–7
Milium effusum L. (*Poaceae*)	H VH SFS R RPL	- 	- 	75–78 1–8 10–13 4–5 1–3	75–78 1–8 10–13 4–5 1–4	75–78 1–8 10–13 4–5 1–5
Miscanthus sacchariflorus (Maxim.) Benth. (*Poaceae*)	H SFS RPL	66–74 9–14 1–4	68–76 9–14 1–5	69–76 8–15 1–5	58–85 6–16 1–7	58–85 6–16 1–7
Naumburgia thyrsiflora (L.) Reichenb. (*Primulaceae*)	H SFS RPL	75–90 9–10 1–2	73–92 9–10 1–3	65–99 8–13 1–4	**61–104** 5–14 1–4	61–104 5–14 1–4
Patrinia scabiosifolia Fisch. ex Link. (*Valerianaceae*)	H SFS RPL	57–78 11–13 1–2	57–78 11–13 1–3	55–80 8–13 1–4	53–90 6–15 1–7	53–90 6–15 1–7
Pedicularis resupinata L. (*Scrophulariaceae*)	H SFS RPL	67–74 6–12 1–3	65–77 5–13 1–3	64–80 4–14 1–4	55–87 2–16 1–6	55–87 2–16 1–6

Appendix 1. (Continued)

Species	Scale	Abundance (%)				
		>8 massales (m)	2,5–8 copiosae (c)	0,2–2,5 numerosae (n)	0,1–0,2 pauces (p)	<0,1 solitariae (s)
1	2	3	4	5	6	7
Persicaria hydropiper (L.) Spach (*Polygonaceae*)	H	76–86	74–87	71–90	67–95	67–95
	SFS	12–14	11–15	10–16	9–17	9–17
	RPL	1–2	1–3	1–4	1–6	1–6
Petasites amplus Kitam. (*Asteraceae*)	H	60–73	60–74	60–75	60–75	60–75
	SFS	9–10	8–10	8–10	8–10	8–10
	RPL	1–2	1–3	1–4	1–4	1–4
Phalaroides arundinacea (L.) Rausch. (*Poaceae*)	H	69–90	60–100	60–100	60–103	60–103
	SFS	9–14	9–14	9–14	8–14	8–14
	RPL	1–4	1–4	1–5	1–6	1–6
Phleum alpinum L. (*Poaceae*)	H	67-69	67-69	65-71	61-72	61-72
	SFS	8-9	8-9	8-9	8-10	8-10
	RPL	1-2	1-3	1-4	1-5	1-5
+*Phleum pratense* L.	H	62–74	60–75	59–76	56–80	56–80
	SFS	10–16	10–16	9–17	7–19	7–19
	RPL	1–3	1–5	1–7	1–8	1–9
Phragmites australis (Cav.) Trin. ex Steud. (*Poaceae*)	H	62–100	62–101	58–101	58–104	58–104
	SFS	10–18	9–17	7–16	3–16	3–16
	RPL	1–2	1–3	1–4	1–6	1–6
Picris japonica Thunb. (*Asteraceae*)	H	-	61–78	61–78	61–78	61–78
	SFS		9–11	9–11	9–11	8–12
	RPL		1–2	1–3	1–4	1–5
Plantago camtschatica Link (*Plantaginaceae*)	H	65–72	64–73	64–74	63–75	63–76
	VH	4–6	4–7	4–8	3–9	2–10
	D	5–6	5–7	5–9	4–10	4–12
	SFS	13–20	11–20	9–21	7–21	5–22
	G	7–10	7–10	6–11	5–11	5–12
	R	15–18	14–18	13–19	12–19	11–20
	Sh	1–3	1–3	1–3	1–4	1–5
	RPL	1–5	1–6	1–7	1–7	1–8
Platycodon grandiflorus (Jacq.) A. DC (*Campanulaceae*)	H	54–60	54–60	52–62	50–74	60–74
	SFS	10–12	10–12	10–12	7–15	7–15
	RPL	1–3	1–3	1–3	1–3	1–3

Appendix 1

Species	Scale	Abundance (%)				
		>8 massales (m)	2,5-8 copiosae (c)	0,2-2,5 numerosae (n)	0,1-0,2 pauces (p)	<0,1 solitariae (s)
1	2	3	4	5	6	7
Poa angustifolia L. (*Poaceae*)	H	63–67	62–69	60–70	59–71	51–72
	VH	8–12	7–13	6–14	5–15	4–16
	D	7–10	7–10	6–11	5–11	5–12
	SFS	7–13	7–13	7–13	6–14	6–14
	G	4–10	4–10	3–12	2–12	1–12
	R	6–14	5–15	4–16	3–17	2–18
	RPL	1–2	1–4	1–5	1–6	1–8
+*Poa annua* L.	H	66–72	64–74	55–82	55–85	39–99
	D	4–11	4–12	3–12	3–12	2–12
	SFS	8–12	7–14	6–15	4–17	2–18
	G	6–12	4–13	3–14	2–14	1–15
	RPL	1–7	1–8	1–8	1–9	1–10
Poa arctica R. Br.	H	65–75	63–78	60–80	58–82	56–84
	VH	7–10	6–11	6–13	5–14	5–16
	D	3–4	2–9	1–10	1–11	1–12
	SFS	8–12	6–14	5–15	4–16	2–18
	G	6–10	5–11	4–12	3–13	2–14
	R	6–10	5–12	5–15	5–17	4–20
	Sh	4–6	3–6	2–7	2–7	1–8
	RPL	1–2	1–4	1–5	1–6	1–9
Poa botryoides (Trin. ex Griseb.) Kom.	H	54–62	54–62	53–67	51–68	51–68
	SFS	8–12	8–12	7–12	7–13	7–13
	RPL	1–2	1–3	1–4	1–4	1–4
Poa glauca Vahl	H	61–69	58–72	58–73	56–74	50–80
	VH	6–11	5–12	3–14	4–13	2–15
	D	7–10	6–11	5–12	5–12	4–13
	SFS	7–11	6–12	3–14	4–13	2–16
	G	8–12	7–13	6–14	7–13	5–15
	R	4–16	3–17	1–19	2–18	1–20
	Sh	3–4	3–5	3–6	3–6	2–8
	RPL	1–4	1–5	1–6	1–6	1–8
Poa kamczatensis Probat.	H	70–74	68–77	67–78	64–80	64–80
	VH	7–8	6–9	6–9	5–10	5–11
	D	9–10	9–10	9–11	9–11	9–12
	SFS	6–8	5–9	4–10	4–11	3–12
	G	9–11	8–12	7–13	6–14	5–15
	R	2–4	2–5	2–5	2–7	2–8
	Sh	1–2	1–3	1–4	1–6	1–7
	RPL	1–3	1–4	1–5	1–6	1–7

Appendix 1. (Continued)

Species	Scale	Abundance (%)				
		>8 massales (m)	2,5–8 copiosae (c)	0,2–2,5 numerosae (n)	0,1–0,2 pauces (p)	<0,1 solitariae (s)
1	2	3	4	5	6	7
Poa macrocalyx Trautv. et C. A. Mey	H	57–86	55–86	54–86	53–86	52–86
	VH	7–11	7–11	6–11	6–12	6–12
	D	5–6	5–6	5–6	4–6	4–10
	SFS	10–18	4–18	3–18	2–18	1–18
	G	7–11	5–11	4–12	3–13	1–14
	R	14–18	14–18	13–19	13–19	12–20
	Sh	1–4	1–4	1–4	1–4	1–4
	RPL	1–2	1–2	1–3	1–4	1–8
Poa malacantha Kom.	H	65–71	63–74	61–76	59–77	57–79
	VH	7–9	7–10	6–12	6–13	5–15
	D	8–12	7–12	7–12	6–12	6–12
	SFS	4–8	3–9	3–10	3–10	1–13
	G	9–13	8–14	8–15	7–15	7–15
	R	1–5	1–7	1–9	1–11	1–13
	RPL	1–2	1–3	1–5	1–6	1–8
Poa nemoralis L.	H	66–70	64–72	63–73	61–74	60–76
	VH	7–8	6–9	4–10	3–11	1–12
	D	5–10	5–10	4–10	4–11	3–12
	SFS	8–10	7–11	6–12	5–14	4–16
	G	4–5	3–6	2–8	3–10	1–11
	R	8–11	7–12	6–12	5–13	4–13
	Sh	8–9	7–9	7–9	6–9	5–10
	RPL	1–3	1–4	1–5	1–6	1–7
Poa neosachalinensis Probat.	H	60-63	58-66	57-69	55-72	53-75
	VH	10-12	9-13	8-14	7-15	6-16
	D	10-12	9-12	9-13	8-14	7-14
	SFS	4-6	3-8	3-9	2-11	1-13
	G	10-12	9-13	8-13	7-15	6-16
	R	3-5	2-8	2-11	1-14	1-20
	Sh	3-4	3-4	3-4	3-4	3-5
	RPL	1-2	1-3	1-4	1-5	1-6
Poa palustris L.	H	8–79	7–80	65–82	62–83	62–84
	D	7–9	6–12	5–12	4–12	2–12
	SFS	10–14	9–15	8–16	7–17	6–18
	G	6–12	6–12	5–13	4–14	3–15
	RPL	2–3	2–4	1–6	1–8	1–9

Appendix 1

Species	Scale	Abundance (%)				
		>8 massales (m)	2,5-8 copiosae (c)	0,2-2,5 numerosae (n)	0,1-0,2 pauces (p)	<0,1 solitariae (s)
1	2	3	4	5	6	7
Poa pratensis L.	H	64–76	61–79	57–83	54–83	50–90
	VH	7–11	6–12	4–14	3–15	3–16
	D	4–7	4–8	3–9	2–11	2–12
	SFS	10–13	9–14	9–15	8–15	7–16
	G	3–5	2–7	2–10	2–12	1–15
	R	7–11	5–12	4–14	2–16	1–17
	Sh	3–4	2–6	2–6	2–8	1–9
	RPL	1–3	1–4	1–5	1–6	1–7
Poa shumushuensis Ohwi	H	-	68–72	68–72	50–80	50–80
	VH		8–10	8–10	5–11	4–12
	SFS		3–6	3–6	2–8	1–9
	RPL		1–4	1–4	1–5	1–6
Poa sibirica Roshev.	H	68–74	66–78	64–83	61–88	59–93
	VH	9–11	8–12	7–13	7–14	6–16
	D	3–5	2–6	2–6	2–6	1–7
	SFS	9–10	8–11	7–11	6–12	5–13
	G	5–6	4–7	3–9	2–10	1–12
	R	11–14	10–15	9–15	8–16	7–17
	Sh	9–10	8–10	7–10	7–10	6–10
	RPL	1–2	1–3	1–4	1–5	1–6
Poa skvortzovii Probat.	H	60–68	58–71	58–73	49–77	49–77
	SFS	5–13	4–14	3–15	1–17	1–17
	RPL	1–3	1–4	1–5	1–8	1–8
Poa stepposa (Kryl.) Roshev.	H	55–57	53–63	49–53	42–69	42–69
	SFS	10–14	10–14	6–10	4–17	4–17
	RPL	3–4	2–5	2–5	1–7	1–7
+ *Poa subcaerulea* Smith	H	77–88	74–89	70–90	67–92	64–93
	VH	4–8	3–9	3–10	2–11	1–12
	D	6–8	5–9	5–9	4–9	4–10
	SFS	6–9	5–11	4–14	4–16	3–19
	G	5–7	5–9	4–10	4–12	3–14
	R	10–11	9–12	8–13	6–14	5–15
	Sh	3–5	3–6	2–6	2–7	1–8
	RPL	1–5	1–6	1–7	1–7	1–8
+ *Poa trivialis* L.	H	67–75	67–75	65–75	62–72	60–75
	VH	1–5	1–5	1–7	1–10	1–12
	SFS	9–11	9–11	9–11	5–14	5–14
	R	1–5	1–5	1–5	1–5	1–5
	RPL	1–3	1–3	1–6	1–8	1–10

Appendix 1. (Continued)

Species	Scale	Abundance (%)				
		>8 massales (m)	2,5-8 copiosae (c)	0,2-2,5 numerosae (n)	0,1-0,2 pauces (p)	<0,1 solitariae (s)
1	2	3	4	5	6	7
Polygonatum humile Fisch. ex Maxim. (*Convallariaceae*)	H SFS RPL	52–67 11–12 1–2	52–67 1–12 1–3	52–67 10–13 1–4	52–67 8–13 1–4	50–70 8–13 1–4
Potentilla anserina L. (*Rosaceae*)	H SFS RPL	70–76 15–17 1–3	69–79 13–17 1–4	67–79 11–17 1–5	65–81 8–18 1–7	65–81 8–18 1–7
Potentilla chinensis Ser.	H SFS RPL	59–65 8–12 3–7	58–67 6–13 3–7	55–69 4–13 2–8	50–74 1–15 1–10	50–74 1–15 1–10
Potentilla fragarioides L.	H SFS RPL	70–75 9–15 1–5	65–80 9–15 1–6	60–85 9–15 1–7	50–95 6–18 1–8	50–95 6–18 1–8
Puccinellia hauptiana V. Krecz. (*Poaceae*)	H D SFS RPL	69–80 2–5 9–17 1–5	68–82 2–6 9–18 1–6	68–84 2–7 9–19 1–7	67–88 1–8 8–20 1–7	67–88 1–9 7–21 1–8
Puccinellia phryganodes (Trin.) Scribn. et Merr.	H D SFS G RPL	91-104 1-5 16-21 5-8 1-6	88-105 1-6 14-21 4-9 1-7	86-106 1-7 10-22 3-10 1-8	83-107 1-8 7-22 3-10 1-8	80-108 1-9 4-22 2-11 1-9
Rabdosia excisa (Maxim.) Hara (*Lamiaceae*)	H SFS RPL	76–70 7–13 1–2	68–78 6–14 1–3	65–80 5–15 1–4	61–84 2–13 1–6	61–84 2–13 1–6
Ranunculus japonicus Thunb. (*Ranunculaceae*)	H SFS RPL	67–73 10–16 1–3	66–74 9–17 1–4	65–75 8–18 1–5	62–77 6–20 1–7	62–77 6–20 1–7
Ranunculus repens L.	H SFS RPL	62–77 10–13 3–5	62–89 9–13 2–7	61–95 8–16 2–7	59–96 7–17 1–8	59–96 7–17 1–8
Reynoutria sachalinensis (Fr. Schmidt) Nakai (*Polygonaceae*)	H SFS RPL	62–74 9–11 1–2	61–75 9–11 1–3	60–75 9–11 1–4	60–75 7–11 1–5	60–75 7–11 1–5

Appendix 1

Species	Scale	Abundance (%)				
		>8 massales (m)	2,5–8 copiosae (c)	0,2–2,5 numerosae (n)	0,1–0,2 pauces (p)	<0,1 solitariae (s)
1	2	3	4	5	6	7
Sanguisorba parviflora (Maxim.) Takeda	H	77–85	74–86	71–90	56–98	56–98
	SFS	9–15	8–16	6–17	4–20	4–20
	RPL	1–3	1–3	1–4	1–5	1–6
Saussurea amurensis Turcz. (*Asteraceae*)	H	67–77	66–78	65–79	62–81	62–81
	SFS	10–14	10–14	9–15	8–16	8–16
	RPL	1–2	1–3	1–3	1–4	1–6
+*Schedonorus pratensis* (Huds.) Beauv. (*Poaceae*)	H	59–73	59–79	59–85	59–86	59–86
	SFS	9–16	9–16	9–16	8–15	8–15
	RPL	1–3	1–4	1–5	1–8	1–8
Schizachne komarovii Roshev. (*Poaceae*)	H	65–71	64–72	63–73	62–74	61–75
	D	5–9	5–10	4–10	4–11	3–11
	SFS	6–12	5–13	5–14	4–14	4–15
	G	4–5	4–6	3–8	2–9	2–11
	RPL	1–5	1–6	1–6	1–7	1–7
Senecio cannabifolius Less. (*Asteraceae*)	H	-	63–75	63–75	63–75	62–76
	SFS		8–12	8–12	7–13	6–14
	RPL		1–2	1–2	1–3	1–3
Senecio pseudoarnica Less.	H	56–83	66–86	65–86	64–87	64–88
	VH	1–2	1–3	1–4	1–5	1–6
	D	6–7	6–8	6–9	5–10	5–12
	SFS	10–19	8–20	7–20	6–21	5–21
	G	12–13	10–13	9–14	7–14	6–15
	R	5–15	4–16	4–16	2–17	1–20
	Sh	1–2	1–3	1–3	1–4	1–5
	RPL	1–3	1–4	1–5	1–6	1–8
+*Setaria pumila* (Poir.) Schult. (*Poaceae*)	H	67–75	63–78	60–82	53–83	53–83
	SFS	10–16	9–17	8–18	5–21	5–21
	RPL	5–7	4–8	3–9	1–10	1–10
Setaria viridis (L.) Beauv.	H	57–65	55–67	53–68	49–73	49-73
	SFS	6–12	5–13	4–14	1–17	1-17
	RPL	4–6	3–7	2–8	1–10	1-10
Silene repens Patr. (*Caryophyllaceae*)	H	53–71	53–71	51–73	44–76	44–76
	SFS	7–12	7–12	5–14	4–14	4–14
	RPL	1–2	1–3	1–4	1–6	1–8
Spodiopogon sibiricus Trin. (*Poaceae*)	H	63–69	60–72	58–75	50–82	50–82
	SFS	6–13	5–19	4–20	2–22	2–22
	RPL	1–3	1–4	1–5	1–8	1–8

Appendix 1. (Continued)

Species	Scale	Abundance (%)				
		>8 massales (m)	2,5-8 copiosae (c)	0,2-2,5 numerosae (n)	0,1-0,2 pauces (p)	<0,1 solitariae (s)
1	2	3	4	5	6	7
Syneilesis aconitifolia (Bunge) Maxim. (*Asteraceae*)	H	57–65	55–67	53–69	49–73	49–73
	SFS	9–13	10–14	8–15	6–16	6–16
	RPL	1–2	1–3	1–3	1–5	1–5
Thermopsis lupinoides (L.) Link (*Fabaceae*)	H	49–65	49–65	47–67	45–74	45–74
	SFS	9–14	9–14	11–16	9–19	9–19
	RPL	2–6	2–6	2–8	2–8	2–8
Trifolium lupinaster L. (*Fabaceae*)	H	57–67	57–67	56–75	53–78	53–78
	SFS	9–13	9–13	8–14	8–15	8–15
	RPL	1–2	1–3	1–4	1–4	1–4
Trifolium pacificum Bobr.	H	63–71	61–73	60–74	56–77	65–77
	SFS	9–13	9–13	8–14	7–15	7–15
	RPL	1–3	1–4	1–5	1–6	1–6
+ *Trifolium pratense* L.	H	61–75	59–77	55–88	56–89	56–89
	SFS	10–13	9–14	9–15	8–17	8–17
	RPL	1–3	1–5	1–8	1–8	1–8
+*Trifolium repens* L.	H	61–77	60–78	59–79	57–80	57–80
	SFS	9–15	9–14	8–16	7–17	7–17
	RPL	3–7	2–8	1–8	1–10	1–10
+*Tripleurospermum inodorum* (L.) Sch. Bip. (*Asteraceae*)	H	62–68	61–69	60–70	57–73	57–73
	SFS	13–15	13–15	12–16	11–16	11–16
	RPL	4–6	4–6	4–6	2–7	1–8
Trisetum molle Kunth. (*Poaceae*)	H	52–82	52–82	47–86	46–88	45–90
	VH	2–12	2–12	1–13	1–13	1–13
	SFS	7–15	5–15	4–15	4–16	3–16
	R	4–8	3–9	3–9	2–9	2–9
	RPL	1–7	1–7	1–8	1–8	1–8
Trisetum sibiricum Rupr.	H	68–70	66–72	65–73	63–75	61–77
	VH	9–11	9–12	8–13	8–14	8–15
	D	5–9	4–10	4–10	3–11	2–12
	SFS	8–11	6–13	4–15	2–17	1–19
	G	5–9	4–10	3–12	2–12	1–13
	Sh	5–6	4–6	3–6	2–7	1–8
	RPL	1–3	1–4	1–5	1–5	1–6
Trisetum spicatum (L.) K. Richt.	H	65–70	64–74	60–75	58–79	55–80
	D	6–9	5–10	4–11	4–11	3–12
	SFS	7–11	6–12	6–12	5–13	4–14
	G	5–12	4–13	3–14	3–14	2–15
	RPL	1–3	1–4	1–5	1–6	1–7

Appendix 1

Species	Scale	Abundance (%)				
		>8 massales (m)	2,5-8 copiosae (c)	0,2-2,5 numerosae (n)	0,1-0,2 pauces (p)	<0,1 solitariae (s)
1	2	3	4	5	6	7
Trollius chinensis Bunge (*Ranunculaceae*)	H SFS RPL	58–82 8–11 1–2	58–82 8–11 1–3	58–82 8–13 1–3	57–82 8–13 1–4	57–82 8–13 1–4
Truellum thunbergii (Siebold. et Zucc) Soják (*Polygonaceae*)	H SFS RPL	- 	- 	61–80 8–11 1–3	61–80 8–11 1–3	61–80 8–11 1–3
Turczaninowia fastigiata (Fisch.) DC. (*Asteraceae*)	H SFS RPL	70–76 10–12 1–3	68–77 10–12 1–4	67–78 9–13 1–5	64–82 7–15 1–7	64–82 7–15 1–7
Urtica angustifolia Fisch. ex Hornem. (*Urticaceae*)	H SFS RPL	64–76 8–11 1–2	64–76 8–11 1–2	64–76 8–11 1–3	60–83 8–12 1–3	60–83 8–12 1–4
Urtica platyphylla Wedd.	H SFS RPL	67–75 8–11 1–2	67–75 8–11 1–2	64–75 8–11 1–3	64–78 8–15 1–3	64–78 8–15 1–4
Veratrum maackii Regel (*Melanthiaceae*)	H SFS RPL	58–65 9–11 1–2	58–65 9–11 1–3	56–67 10–12 1–3	56–68 9–12 1–3	56–68 9–12 1–3
Vicia cracca L. (*Fabaceae*)	H SFS RPL	67–75 9–15 1–3	63–79 8–16 1–5	58–84 7–17 1–6	49–93 5–19 1–7	49–93 5–19 1–7
Vicia unijuga A. Br.	H SFS RPL	66–74 5–13 1–3	64–75 4–14 1–3	62–77 3–14 1–4	58–79 2–15 1–6	58–79 2–15 1–6
Viola acuminata Ledeb. (*Violaceae*)	H SFS RPL	61–71 9–10 1–2	61–71 9–10 1–2	59–73 9–10 1–3	57–75 7–12 1–5	57–75 7–12 1–5
Viola patrinii Ging.	H SFS RPL	70–80 11–15 1–3	66–83 10–16 1–4	63–86 9–17 1–5	55–94 7–19 1–5	55–94 7–19 1–5
Zizania latifolia (Griseb.) Stapf (*Poaceae*)	H SFS RPL	100–105 12–14 1–2	101–106 12–14 1–2	102–107 12–15 1–2	104–109 10–16 1–3	104–109 10–16 1–3

APPENDIX 2.

Ecological ranges for the representatives of the Family *Poaceae*.

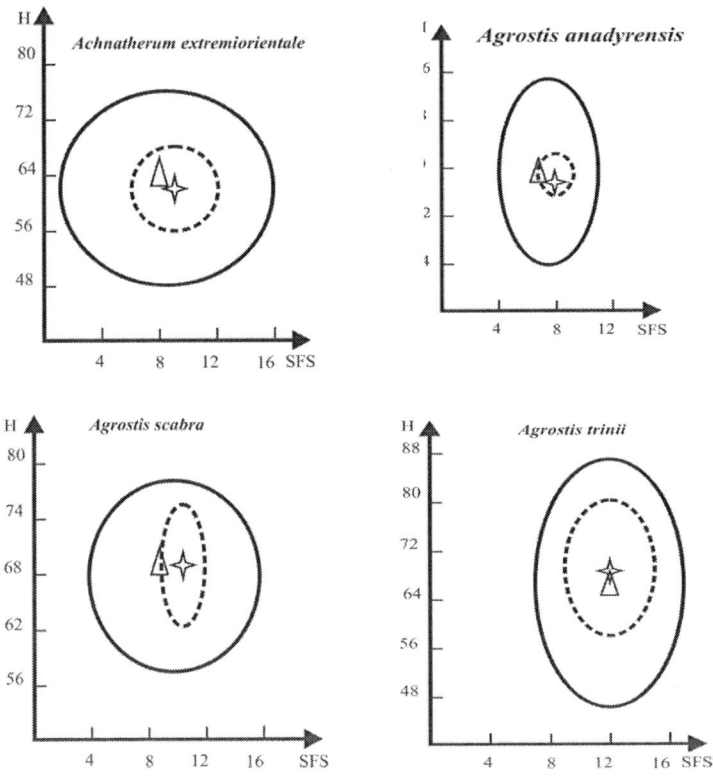

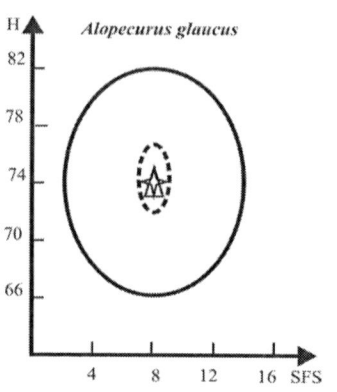

Alopecurus glaucus

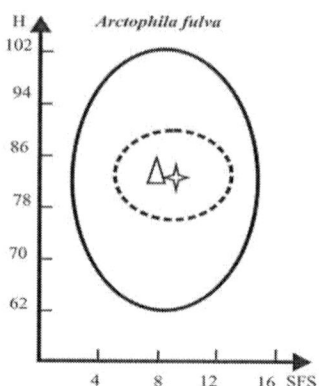

Arctophila fulva

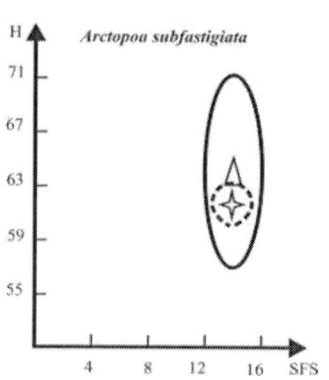

Arctopoa subfastigiata

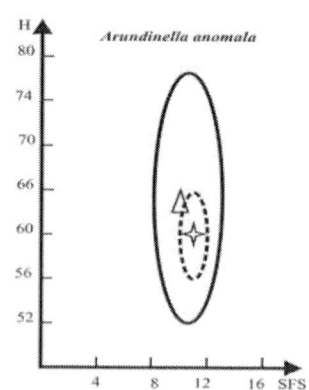

Arundinella anomala

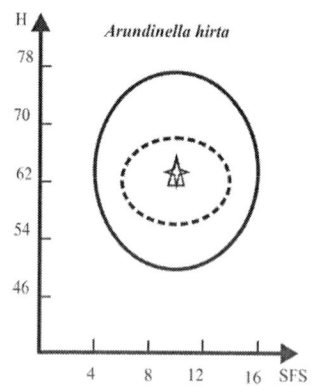

Arundinella hirta

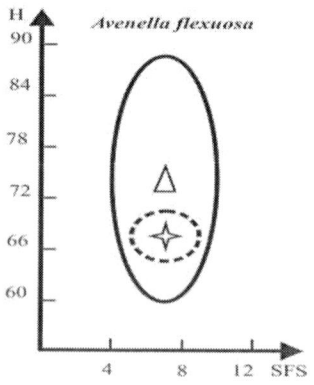

Avenella flexuosa

Appendix 2

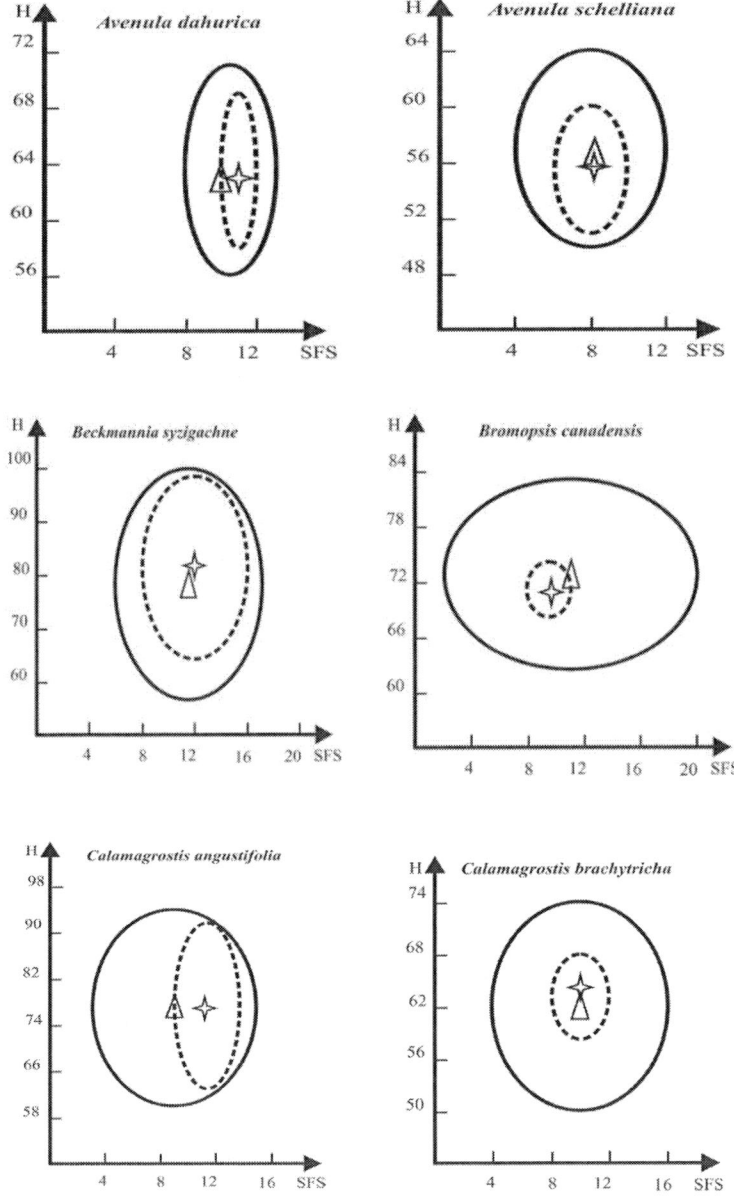

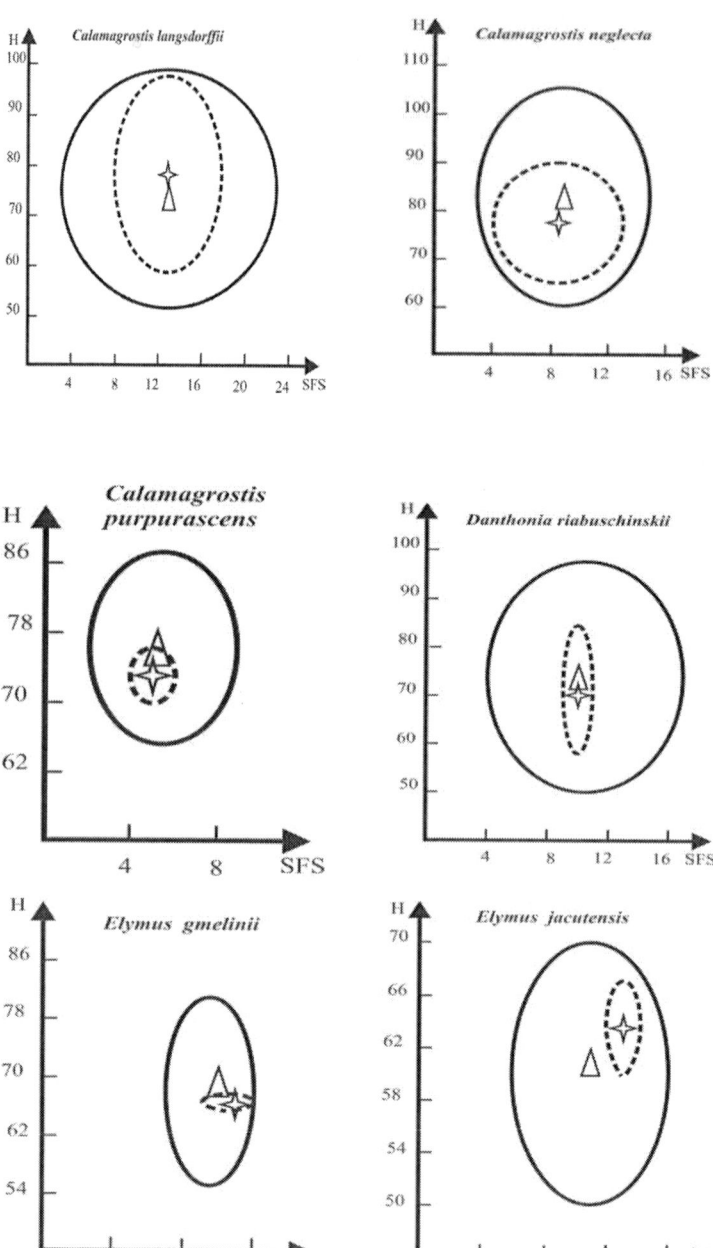

Appendix 2

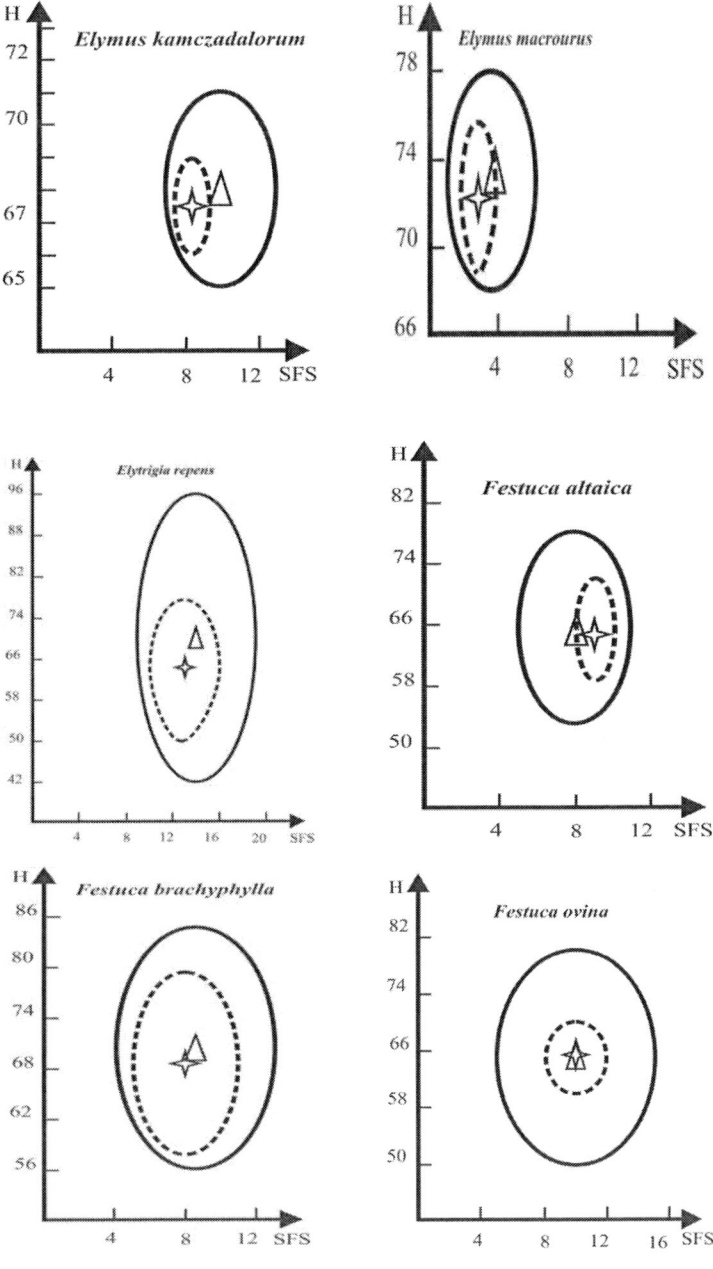

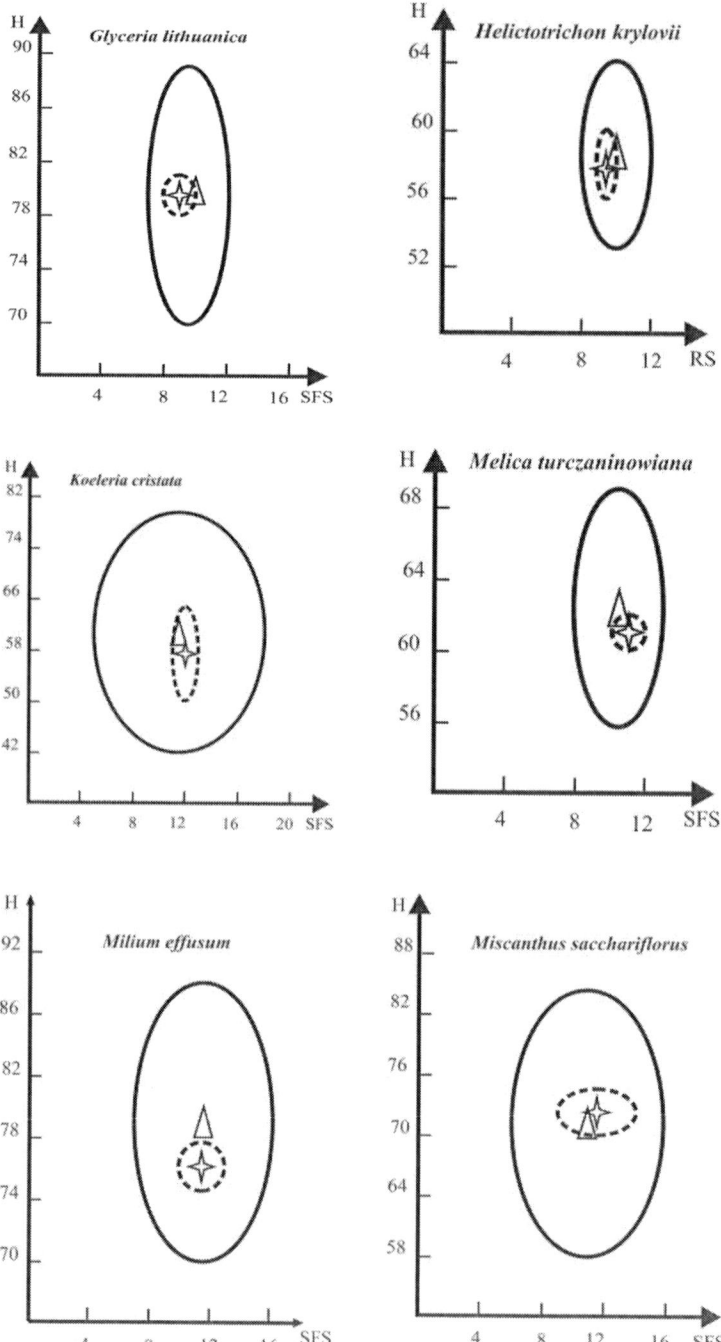

Appendix 2

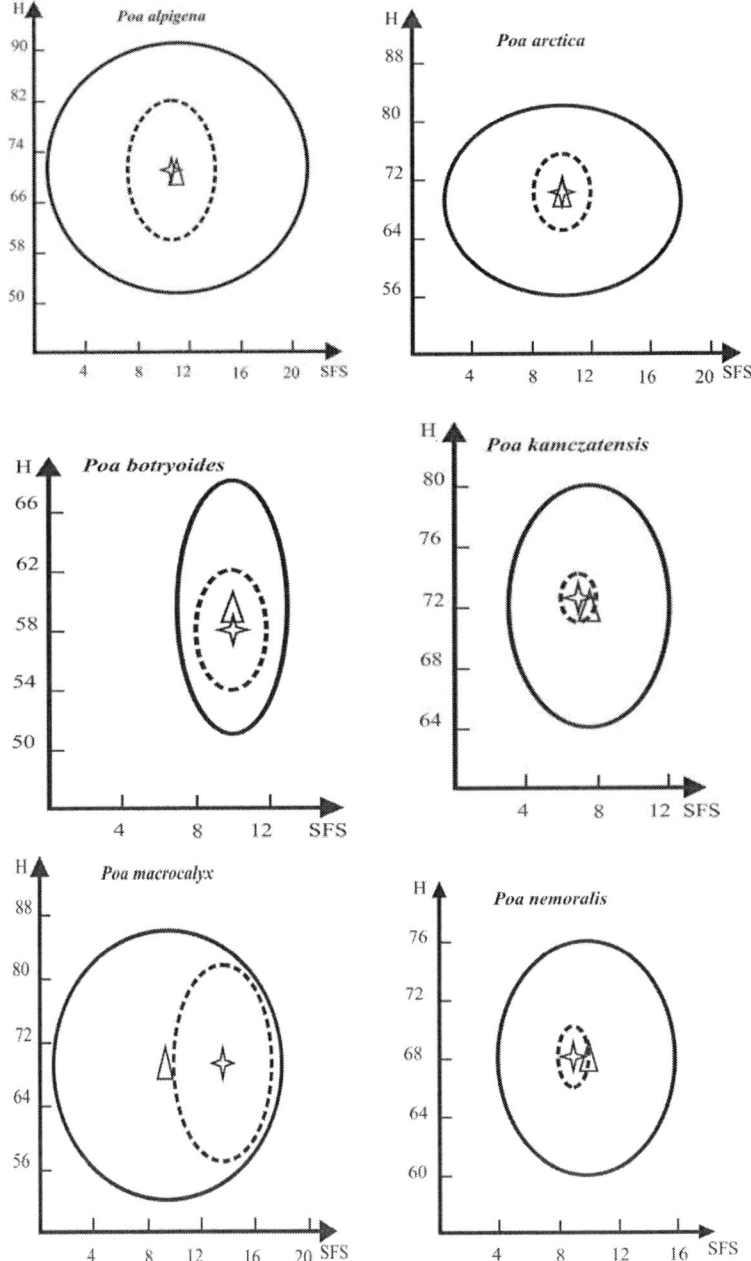

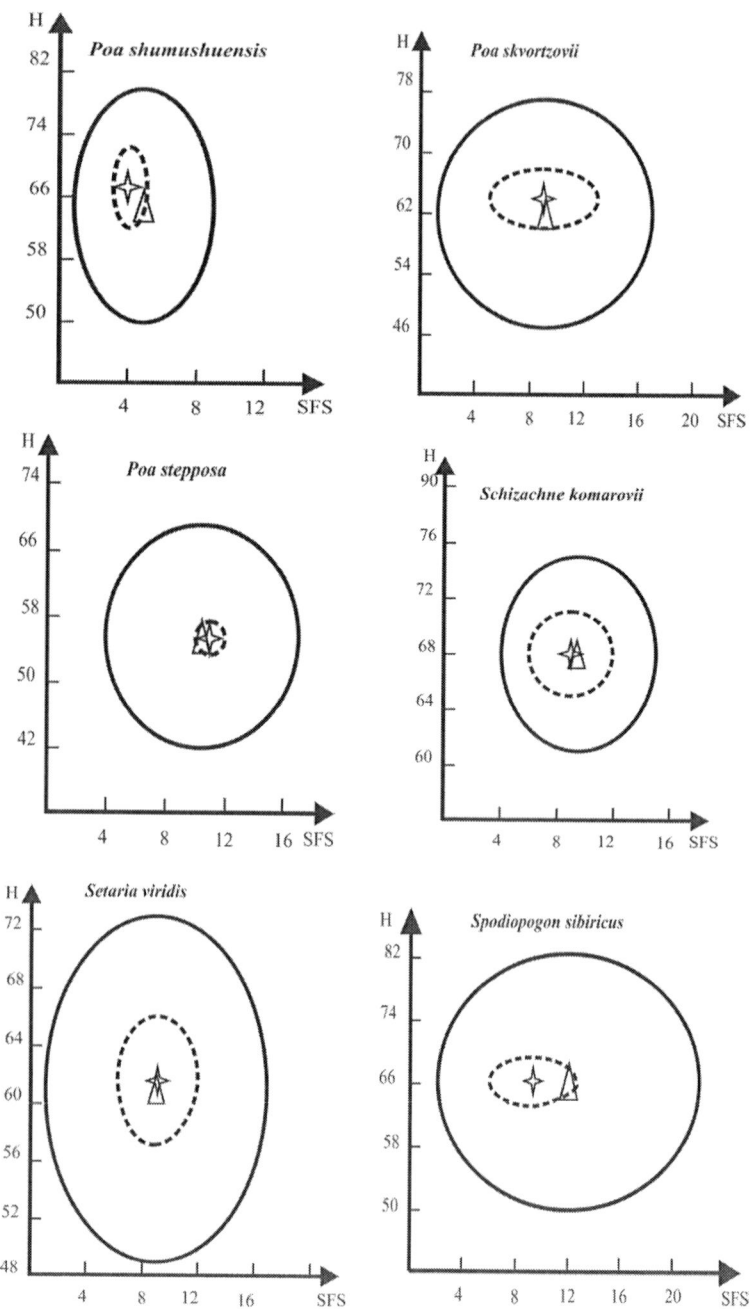

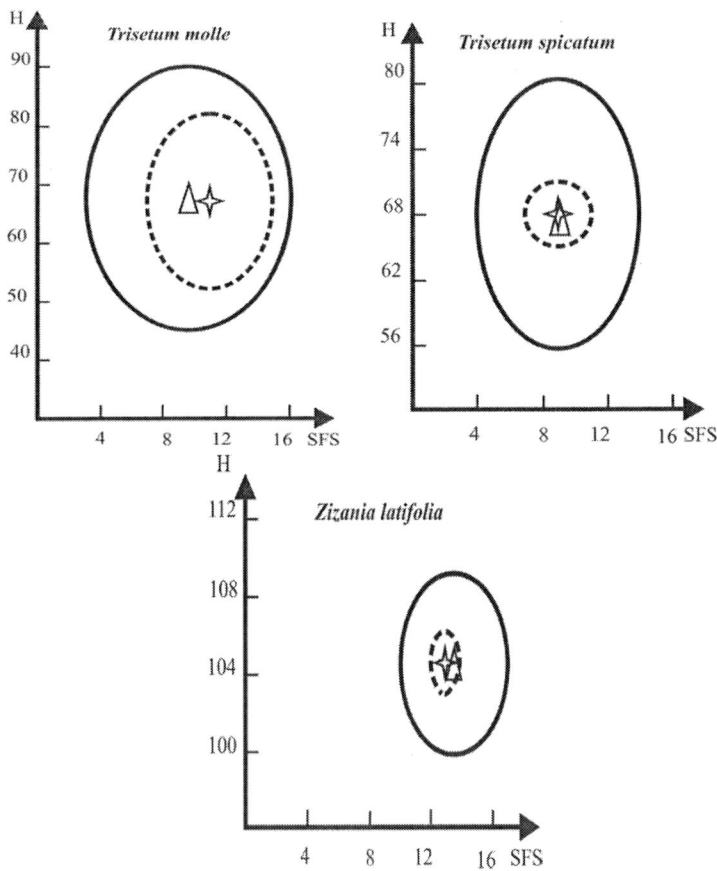

ABOUT THE AUTHORS

Vitaly Pavlovich Seledets, DSci.

Pacific Institute of Geography, Far East Branch of the Russian Academy of Sciences, Vladivostok, Russia

Vitaly P. Seledets graduated from the Rostov-on-Don State University in 1962. His fields of study include plant ecology, phytogeography, and nature conservation. His scientific interests include ecology of plant communities, ecological ranges and ecological niches, plants in the "land - ocean" contact zone, problems of the wildlife protection. He is the lead researcher at the Laboratory of Biogeography and Ecology, Pacific Institute of Geography, Far East Branch of the Russian Academy of Sciences (PIG FEB RAS). He has lectured as a professor (botany, plant ecology, phytogeography, social ecology, and nature conservation) at the Far East State University, the Vladivostok University of Economics and Service, Pacific University of Economics. Dr. Seledets has authored over 280 scientific publications, among them the monographs "Nature protection complexes of the Far East" (1991), "Protected wildlife areas of Primorsky Territory" (1993), "Ecological scales for vegetation studies in the Russian Far East" (2000), "Anthropogenic dynamics of vegetation of the Russian Far East" (2000), "Vegetation of the nature monuments in the islands of the Peter the Great Bay (Primorsky Territory)" (2000), "Vegetation of the nature monuments in the coastal zone of the Sea of Japan (Primorsky Territory)" (2005), and "Vegetation of the nature monuments in the coastal zone of the Peter the Great Bay (southwest part of Primorsky Territory)" (2009). He was the main author of "Vegetation and tourism" (2000) and "Ecological range of plant species" (2007). He has made many field trips to various regions of the Russian Far East, Yakutia and other regions of Russia as well as the former USSR. He is included in "Who is who in the world" for 2011.

Nina Serguejevna Probatova, DSci.

Institute of Biology and Soil Science, Far East Branch of the Russian Academy of Sciences, Vladivostok, Russia

Nina S. Probatova graduated from the Rostov-on-Don State University in 1961. Her fields of study include vascular plant taxonomy and karyosystematics, phytogeography, and floristics. Her main scientific interests are taxonomy and karyosystematics of the *Poaceae* Family, chromosome numbers as a source of information on the flora of the Russian Far East, and problems of sea coastal botany. She is the principal researcher at the Laboratory of Vascular Plants, Institute of Biology & Soil Science, Far East Branch of the Russian Academy of Sciences (IBSS FEB RAS). Dr. Probatova has over 290 scientific publications and was one of the main authors of the "Vascular Plants of the Soviet Far East", vols. 1-8 (1985-1996): *Poaceae, Lamiaceae* and some other Dicot families, the first author of "Karyology of the flora of Sakhalin and the Kuril Islands" (2007), and "Flora of the Russian Far East. Addenda et Corrigenda ..." (2006). She was also one of the authors of "Manual of higher plants of Sakhalin and the Kurils" (1974), "Manual of the vascular plants of the Kamchatskaya Oblast" (1981), and "Wild fodder grasses of the Far East" (1982). Dr. Probatova has authored publications on chromosome numbers and plant karyotaxonomy since 1968. She has led projects sponsored by the Russian Fund for Basic Researches (RFBR), including "Karyology of the flora in the southern part of the Russian Far East" (1998-1999), "The Grass Family (*Poaceae)* in the flora of the Russian Far East: biodiversity, biogeography, and evolution " (2001-2002), "Vascular plants in the "land - ocean" contact zone [Russian Far East]: biodiversity, ecology, and phytogeography" (2004-2005), and now - "The Grass Family (*Poaceae)* of Russia: taxonomic revision, phylogeny, karyosystematics, geography" (2011-2013). From the Far East of Russia (and from other regions of the former USSR and some other territories), she described new taxa of *Agrostis, Arctopoa, Calamagrostis, Calystegia, Clinopodium, Deschampsia, Dianthus, Dracocephalum, Elsholtzia, Eragrostis, Elymus, Glyceria, Hierochloë, Lycopus, Lysimachia, Melica, Milium, Papaver, Poa, Thymus, Trisetum,* and *Zingeriopsis*. She has made many field trips, since 1959 to Sakhalin, the Kurils, Kamchatka, North Koryakia, the Commander Islands, the Kolyma River basin, the Sea of Okhotsk coast, the Amur River basin, and Primorsky Territory (especially the Sea of Japan continental coast and the islands of Peter the Great Bay). She is one of the authors of "Red Data Book of the Sakhalinskaya Oblast" (2005), "Red Data Book of Primorsky Territory" (2008), and "Red Data Book of the Russian Federation" (2008).

INDEX

A

Achnatherum sibiricum, 46, 111
Aconitum albo-violaceum, 65
adaptation, xii, 15, 19, 27, 29, 31, 39, 54, 58, 70, 73, 77
Adenophora pereskiifolia, 65
Agrostis alaskana, 12, 87
Agrostis anadyrensis Socz., 43, 112
Agrostis capillaris L., 112
Agrostis clavata Trin., 44, 112
Agrostis gigantea Roth, 112
Agrostis scabra Willd., 44, 112
Agrostis sokolovskajae, xii, 13
Agrostis stolonifera L., 113
Agrostis trinii Turcz., 47, 113
Aizopsis aizoon, 65
Aizopsis maximowiczii, 8
Ajania pallasiana, 65
Allium senescens, 65
Alopecurus aequalis Sobol., 44
Alopecurus arundinaceus, 52, 53
Alopecurus brachystachyus Bieb., 45
Alopecurus glaucus Less., 43, 113
Alopecurus pratensis, 53
amplitude, 34, 57, 71
Anthoxanthum odoratum, 52
anthropotolerance, 19, 20

Arctagrostis latifolia (R. Br.) Griseb., 113
Arctophila fulva (Trin.) Anderss., 45, 113
Arctopoa eminens (J. S. Presl) Probat., 48, 113
Arctopoa subfastigiata (Trin.) Probat., 114
Artemisia gmelinii, 3, 65, 67
Artemisia keiskeana Miq., 114
Artemisia mandshurica (Kom.) Kom., 114
Artemisia stelleriana, 8
Artemisia stolonifera (Maxim.) Kom., 114
Arundinella anomala Steud., 48, 114
Arundinella hirta (Thunb.) Tanaka, 48, 114
Aster maackii, 34
Astragalus marinus, 8, 9
Atriplex patens, 52
Atriplex subcordata, 9
Avenula dahurica (Kom.) Holub, 114
Avenula schelliana (Hack.) Holub, 47, 114

B

Beckmannia syzigachne (Steud.) Fern., 45, 115
belt, 1, 2, 4, 7, 20, 48, 50, 63, 67, 70, 71, 84
  maritimal, 1
  maritime, 2, 4, 6, 20, 48, 50, 67, 70, 71
botany, xii, 1, 2, 78, 86, 91, 95, 143, 145
  coastal, 1, 2, 3, 4, 6, 7, 8, 9, 10, 13, 15, 19, 29, 31, 34, 39, 40, 41, 48, 55, 58, 62, 63, 67, 71, 77, 79, 82, 84, 91, 92, 95, 97, 144
Bothriospermum tenellum, xii
Bromopsis canadensis (Michx.) Holub, 42, 115
Bromopsis inermis (Leys.) Holub, 115
Bromopsis pumpelliana (Scribn.) Holub, 115
Bupleurum longiradiatum Turcz., 115

C

Cakile edentula, 78, 99
Calamagrostis amurensis, 14
Calamagrostis angustifolia Kom., 45, 115
Calamagrostis brachytricha Steud., 42, 115
Calamagrostis deschampsioides Trin., 48
Calamagrostis epigeios, 53
Calamagrostis langsdorffii (Link) Trin., 116
Calamagrostis litwinowii, 12
Calamagrostis neglecta (Ehrh.) Gaertn., Mey. et Schreb., 116
Calamagrostis purpurascens R. Br., 46
Calamagrostis sachalinensis Fr. Schmidt, 116
Calamagrostis sesquiflora (Trin.) Tzvel., 46, 116
Calystegia soldanella, 5

Carex appendiculata (Trautv. et Mey.) Kiik., 117
Carex bostrichostigma, 68
Carex campylorhina V. Krecz., 117
Carex falcata Turcz., 117
Carex kobomugi, 6
Carex macrocephala, 6, 9
Carex nanella Ohwi, 117
Carex pallida C.A. Mey., 117
Carex schmidtii Meinsh., 117
Carex siderosticta Hance, 117
Carex ussuriensis Kom., 117
Chelidonium asiaticum, 6
Chloranthus japonicus Siebold, 117
Chorisis repens, 5, 8, 9
chromosome, ix, xiii, 2, 4, 6, 7, 13, 14, 34, 63, 82, 85, 88, 89, 90, 91, 92, 100, 144
  counts, xiii
  numbers, xiii, 2, 4, 6, 7, 14, 63, 82, 85, 86, 87, 88, 89, 90, 91, 92, 93, 100, 144
  studies, ix, xiii, 7, 9, 11, 16, 17, 18, 20, 33, 34, 55, 63, 73, 76, 81, 87, 88, 89, 90, 91, 92, 95, 96, 97, 99, 144
Cinna latifolia, 12
Cirsium coryletorum, 7
Cleistogenes kitagawae Honda, 47, 117
Clematis hexapetala, 65
Clinopodium chinense, 68
coenogenesis, xii
coenopopulation, 16, 20, 21, 49, 80
Coleanthus subtilis, xii
community, vii, 2, 20, 25, 30, 100
  coastal, 1, 2, 3, 4, 6, 7, 8, 9, 10, 13, 15, 19, 29, 31, 34, 39, 40, 41, 48, 55, 58, 62, 63, 67, 71, 77, 79, 82, 84, 91, 92, 95, 97, 144
  plant, vii, xi, xii, xiii, 1, 2, 3, 4, 7, 8, 11, 13, 15, 16, 18, 19, 20, 21, 22, 23, 25, 29, 30, 31, 32, 33, 34, 37, 39, 40, 41, 44, 45, 47, 48, 49, 50, 51, 53, 55, 58, 63, 64, 65, 67, 70, 73, 76, 77, 78, 79, 80, 83, 85, 87,

88, 89, 91, 92, 94, 95, 96, 97, 98, 99, 103, 111, 143, 144
concept, vii, 16, 17, 18, 19, 20, 21, 23, 61, 79, 83, 97, 99
  ecological range, vii, 3, 11, 13, 19, 20, 23, 53, 61, 73, 80, 97, 98, 143
  individualistic, 17, 79, 82, 83

D

Dactylis glomerata L., 118
Danthonia riabuschinskii (Kom.) Kom., 48, 118
Dasiphora mandshurica, 65
Deschampsia macrothyrsa, 9
Dianthus amurensis, 68
Dianthus stepanovae, 9
differentiation, 2, 19, 24, 35, 37, 38, 39, 54, 63, 71, 76, 88, 94, 97, 99
  floristic, 2, 5, 6, 9, 10, 11, 15, 34, 51, 65, 67, 82, 101
  taxonomic, xii, xiii, 5, 12, 13, 15, 23, 25, 28, 37, 38, 39, 63, 76, 81, 86, 87, 89, 101, 145
Dontostemon dentatus, 66
drainage, 19, 20, 24, 46, 55, 71, 111

E

ecological, vii, xii, xiii, 1, 2, 3, 5, 8, 9, 10, 11, 13, 15, 16, 17, 18, 19, 20, 21, 23, 24, 25, 26, 27, 28, 29, 30, 31, 33, 34, 37, 38, 39, 40, 41, 42, 45, 47, 48, 49, 50, 52, 53, 55, 57, 58, 61, 62, 63, 64, 65, 67, 69, 70, 71, 73, 76, 79, 80, 81, 82, 91, 94, 95, 96, 97, 98, 99, 101, 102, 143
ecotone, xi, 39
Elymus confusus (Roshev.) Tzvel., 46, 118
Elymus gmelinii (Ledeb.) Tzvel., 42, 118
Elymus jacutensis (Drob.) Tzvel., 44, 118
Elymus kamczadalorum (Nevski) Tzvel., 42, 118
Elymus kronokensis (Kom.) Tzvel., 118
Elymus macrourus (Turcz.) Tzvel., 118
Elymus sibiricus L., 119
Elymus woroschilowii, 8, 10
Elytrigia repens (L.) Nevski, 119
environment, xi, 1, 2, 8, 23, 29, 79
  coastal, 1, 2, 3, 4, 6, 7, 8, 9, 10, 13, 15, 19, 29, 31, 34, 39, 40, 41, 48, 55, 58, 62, 63, 67, 71, 77, 79, 82, 84, 91, 92, 95, 97, 144
ephemer
  hydrophilous, xii
Eriochloa villosa, 66
Eupatorium lindleyanum DC., 119
Euphrasia maximowiczii, 68
evaluation, 15, 16, 18, 20, 94, 97, 101

F

factor, xii, 9, 19
  climatic, 11, 12, 73
  environmental, xiii, 1, 2, 3, 8, 15, 16, 17, 18, 19, 20, 21, 23, 25, 26, 34, 39, 54, 56, 58, 70, 71, 73
Festuca altaica Trin., 46, 119
Festuca amurensis, 13
Festuca brachyphylla Schult. et Schult. fil., 119
Festuca extremiorientalis Ohwi, 42, 119
Festuca hondoensis, 12
Festuca limosa, 12
Festuca ovina L., 119
Festuca rubra L., 119
Filipendula camtschatica (Pall.) Maxim., 119
Filipendula palmata (Pall.) Maxim., 119
Fimbripetalum radians (L.) Ikonn., 119
flora, xi, xiii, 2, 3, 4, 5, 6, 7, 9, 11, 12, 13, 21, 34, 37, 52, 53, 63, 75, 76, 77, 78, 79, 81, 82, 85, 86, 87, 88, 89, 90, 91, 92, 94, 95, 98, 101, 102, 103, 144
  continental, xi, 3, 7, 10, 11, 12, 16, 39, 41, 45, 47, 48, 49, 50, 52, 53,

55, 56, 57, 58, 59, 60, 61, 73, 84, 97, 98, 145
sea coastal, 3, 21, 52, 84, 95, 145
vascular, xiii, 4, 6, 7, 9, 11, 17, 29, 34, 77, 86, 87, 88, 89, 90, 91, 92, 93, 96, 97, 100, 111, 144
florogenesis, vii, xii, 1, 15
Fragaria orientalis Losinsk., 119

G

Galium boreale L., 120
Galium davuricum, 68
Galium verum L., 120
Geranium erianthum DC., 120
Geranium eriostemon Fisch., 120
Geranium sibiricum L., 120
Geranium wlassowianum Fisch. ex Link, 120
Geum aleppicum Jacq., 120
Glaux maritima, 5
Glehnia littoralis, 4, 5, 7, 8, 9
Glyceria lithuanica (Gorski) Gorski, 120
Glyceria spiculosa (Fr. Schmidt) Roshev., 120
Glyceria triflora (Korsh.) Kom., 120
Glyceria voroschilovii, 13
Glycine soja Siebold et Zucc., 121
Gypsophila pacifica Kom., 121

H

habitat, 10, 31, 94
　coastal, 1, 2, 3, 4, 6, 7, 8, 9, 10, 13, 15, 19, 29, 31, 34, 39, 40, 41, 48, 55, 58, 62, 63, 67, 71, 77, 79, 82, 84, 91, 92, 95, 97, 144
　seaside, 1, 4, 6, 8, 19, 39
　xerophytic, 2, 6
halophyte, 5
　coastal, 1, 2, 3, 4, 6, 7, 8, 9, 10, 13, 15, 19, 29, 31, 34, 39, 40, 41, 48, 55, 58, 62, 63, 67, 71, 77, 79, 82, 84, 91, 92, 95, 97, 144

facultative, 4, 5
obligate, 4, 5, 13
Heteropappus hispidus, 66
Heteropappus saxomarinus, 7, 9
Hieracium umbellatum L., 121
Hierochloë alpina (Sw.) Roem. et Schult., 121
Hierochloë glabra, 14, 29
Hierochloë kamtschatica, 12
Hierochloë pauciflora R.Br., 45
Honckenya oblongifolia Torr. et Gray, 121
Hordeum brachyantherum, 12
Hordeum jubatum, 4, 52
Hordeum roshevitzii, 6
humidity, 1, 19, 20, 21, 23, 37, 38, 40, 43, 45, 46, 52, 55, 57, 58, 65, 67, 70, 71, 111
hybrid, xiii, 2, 10, 13, 37, 77, 78
hybridization, 13
Hypericum attenuatum, 66, 68
Hypericum gebleri Ledeb., 121

I

Iris ensata Thunb., 121
Iris setosa Pall. ex Link, 121
Iris uniflora, 66

J

Juncus bufonius L., 121
Juniperus davuricus, 66

K

Kalimeris incisa (Fisch.) DC., 121
Kitagawia litoralis, 5
Kitagawia terebinthacea, 66
Koeleria ascoldensis, 6, 8, 9, 10
Koeleria cristata (L.) Pers., 47, 122

Index

L

Lamium barbatum, 34
Lathyrus japonicus Willd., 122
Lathyrus quinquenervius, 66
Leontopodium leontopodioides, 66
Lespedeza bicolor, 66
Lespedeza juncea (L. fil.) Pers., 122
level, xi, 2, 9, 12, 19, 34, 51
    diploid, 9, 33, 34
    local, 20
    ploidy, xiii, 2, 5, 7, 9, 14, 33, 34, 50
    regional, 18, 19, 21, 52, 53, 80
    subregional, 19
    tetraploid, 50
Leymus chinensis (Trin.) Tzvel., 122
Leymus coreanus, 68
Leymus mollis (Trin.) Hara, 122
Ligusticum hultenii Fern., 122
Lilium buschianum, 66
Lilium pensylvanicum Ker-Gawl., 122
Limonium tetragonum, 84
Linaria vulgaris Mill., 123
load, 20, 24, 55, 58, 65, 67, 76, 78, 111
    pasture, 20, 24, 55, 58, 65, 67, 111
    recreational, 20, 24, 55, 78, 111
Lysimachia davurica Ledeb., 123
Lythrum salicaria L., 123

M

Macrohystrix komarovii, 12
Malus mandshurica, 3
Melica turczaninowiana Ohwi, 47
Mertensia simplicissima (Ledeb.) G. Don fil., 123
Milium effusum L., 123
Miscanthus sacchariflorus (Maxim.) Benth., 48, 123
Miscanthus sinensis, 68

N

Naumburgia thyrsiflora (L.) Reichenb., 123
Neomolinia japonica, 12
niche, vii, 24, 25, 33, 55, 61, 70, 73, 79, 80, 84, 94

O

optimum, 5, 25, 29, 30, 31, 45, 52, 53, 73
Orostachys malacophylla, 66, 68

P

Paeonia obovata, 66
Papaver sokolovskajae, 8
Paraixeris denticulata, 7
Patrinia rupestris, 66
Patrinia scabiosifolia Fisch. ex Link., 123
Pedicularis resupinata L., 123
Peracarpa circaeoides, 7
Persicaria hydropiper (L.) Spach, 124
Petasites amplus Kitam., 124
Phalaroides arundinacea (L.) Rausch., 43, 124
Phleum alpinum L., 124
Phleum pratense L., 124
Phragmites australis (Cav.) Trin. ex Steud., 124
Phryma asiatica, 34
Physocarpus amurensis, 66
phytoindication, 16, 19, 20, 77
Picris japonica Thunb., 124
plant, vii, xi, xii, xiii, 1, 2, 3, 4, 7, 8, 11, 13, 15, 16, 18, 19, 20, 21, 22, 23, 25, 29, 30, 31, 32, 33, 34, 37, 39, 40, 41, 44, 45, 47, 48, 49, 50, 51, 53, 55, 58, 63, 64, 65, 67, 70, 73, 76, 77, 78, 79, 80, 83, 85, 87, 88, 89, 91, 92, 94, 95, 96, 97, 98, 99, 103, 111, 143, 144
    alien, 11, 13, 51, 52, 53, 80, 111

coastal, 1, 2; 3, 4, 6, 7, 8, 9, 10, 13,
15, 19, 29, 31, 34, 39, 40, 41, 48,
55, 58, 62, 63, 67, 71, 77, 79, 82,
84, 91, 92, 95, 97, 144
vascular, xiii, 4, 6, 7, 9, 11, 17, 29,
34, 77, 86, 87, 88, 89, 90, 91, 92,
93, 96, 97, 100, 111, 144
Plantago camtschatica Link, 124
Plantago depressa, 68
ploidy
level, xiii, 2, 5, 7, 9, 14, 33, 34, 50
Poa almasovii, 9, 10, 13, 87
Poa alpigena (Blytt.) Lindm., 43
Poa angustifolia L., 125
Poa annua L., 125
Poa arctica R. Br., 125
Poa botryoides (Trin. ex Griseb.) Kom., 125
Poa glauca Vahl, 125
Poa kamczatensis Probat., 47, 125
Poa macrocalyx Trautv. et C. A. Mey, 48, 126
Poa malacantha Kom., 126
Poa nemoralis L., 42, 126
Poa neosachalinensis Probat., 46, 126
Poa palustris L., 43, 126
Poa pratensis L., 127
Poa shumushuensis Ohwi, 44, 127
Poa sibirica Roshev., 127
Poa sichotensis, 13
Poa skvortzovii Probat., 42, 127
Poa stepposa (Kryl.) Roshev., 46, 127
Poa subcaerulea Smith, 127
Poa tatewakiana, 34
Poa trivialis L., 127
Poa turneri, 12
Polygonatum humile Fisch. ex Maxim., 128
polymorphism, xiii, 8, 14
  intraspecific, 8, 14
  karyological, xiii, 7, 15
polyploid, 7, 76
population, vii, xii, 1, 2, 9, 20, 39, 55, 76, 78, 80, 99, 101

coenotic, 2, 3, 20, 21, 29, 83, 97, 98, 103
structure, vii, xii, 1, 2, 8, 10, 23, 37, 49, 52, 55, 64, 78, 79, 81, 97, 101, 103
Potentilla anserina L., 128
Potentilla chinensis Ser., 128
Potentilla fragarioides L., 128
Primula cuneifolia, 7
profile, 57, 58, 73
geographical, vii, xii, 6, 7, 11, 16, 20, 21, 23, 33, 34, 39, 41, 49, 56, 57, 58, 59, 60, 61, 63, 71, 73, 85, 87, 88, 90, 96, 98
psammophyte
coastal, 1, 2, 3, 4, 6, 7, 8, 9, 10, 13, 15, 19, 29, 31, 34, 39, 40, 41, 48, 55, 58, 62, 63, 67, 71, 77, 79, 82, 84, 91, 92, 95, 97, 144
Puccinellia hauptiana V. Krecz., 128
Puccinellia kurilensis, 5, 9
Puccinellia nipponica, 4, 8, 12
Puccinellia phryganodes (Trin.) Scribn. et Merr., 48, 128

Q

Quercus mongolica, 3

R

Rabdosia excisa (Maxim.) Hara, 128
range, vii, 4, 19, 23, 47, 49, 99, 144
Ranunculus japonicus Thunb., 128
Ranunculus repens L., 128
region, xi, xii, 11, 16, 41, 43, 49, 92
  monsoon, vii, xi, xii, xiii, 1, 3, 6, 8, 9, 13, 15, 16, 18, 19, 21, 34, 39, 41, 43, 50, 51, 58, 61, 64, 72, 73, 75, 76, 79, 83, 84, 85, 88, 90, 91, 93, 94, 97, 99, 103
  Pacific, xi, xii, 5, 6, 7, 9, 12, 13, 16, 18, 24, 34, 37, 38, 39, 40, 41, 42, 43, 44, 45, 46, 47, 48, 49, 50, 51,

55, 57, 58, 61, 63, 64, 72, 73, 76, 79, 87, 94, 96, 97, 98, 143
phytogeographical, xii, 7, 21, 23, 79, 89
relict (relic)
  taxonomic, xii, xiii, 5, 12, 13, 15, 23, 25, 28, 37, 38, 39, 63, 76, 81, 86, 87, 89, 101, 145
  thermophilic, xii
Reynoutria sachalinensis (Fr. Schmidt) Nakai, 128
Rhodiola integrifolia, 7
Rosa rugosa, 3, 5, 6, 9, 10, 108

S

Salsola komarovii, 4, 9
Sanguisorba parviflora (Maxim.) Takeda, 129
Saussurea amurensis Turcz., 129
Saussurea neopulchella, 7
scale, 20, 39, 94
  shading, 19, 20, 55, 71, 111
Schedonorus pratensis (Huds.) Beauv., 129
Schizachne komarovii Roshev., 43, 129
Scirpus lineolatus, 22
Scrophularia grayana, 9
Scutellaria strigillosa, 5, 8
seacoast, 20, 62, 67, 71
seaside, 1, 4, 6, 8, 19, 39
Securinega suffruticosa, 66
Selaginella tamariscina, 66
Senecio cannabifolius Less., 129
Senecio pseudoarnica Less., 129
Serratula mandshurica, 66
Setaria pachystachys, 8, 9, 12
Setaria pumila (Poir.) Schult., 129
Setaria viridis (L.) Beauv., 44, 129
Silene foliosa, 66
Silene repens Patr., 129
soil, 1, 19, 20, 21, 23, 38, 40, 52, 55, 58, 65, 67, 70, 71, 94, 111
  fertility, 19, 20, 21, 23, 38, 40, 52, 55, 58, 65, 67, 70, 71, 111

granulometric composition, 19, 20, 24, 46, 70, 71, 111
renewal, 1, 19, 20, 55, 111
salinity, 4, 19, 20, 21, 23, 38, 40, 52, 55, 58, 65, 67, 70, 71, 111
Sonchus asper, 52
Sophora flavescens, 66
space, vii, 23, 24, 25, 26, 33, 55, 73
  multidimensional, vii, 24, 55, 73
  geographical, vii, xii, 6, 7, 11, 16, 20, 21, 23, 33, 34, 39, 41, 49, 56, 57, 58, 59, 60, 61, 63, 71, 73, 85, 87, 88, 90, 96, 98
speciation, vii, xii, 1, 2, 8, 10, 13, 14, 79, 87
species, vii, xii, xiii, 1, 2, 3, 4, 5, 6, 7, 8, 9, 10, 11, 12, 13, 15, 16, 17, 18, 19, 20, 21, 22, 23, 25, 26, 29, 30, 31, 32, 33, 34, 37, 38, 39, 40, 41, 43, 44, 45, 47, 48, 49, 50, 51, 52, 53, 55, 58, 61, 63, 64, 65, 67, 70, 71, 73, 75, 76, 77, 78, 79, 80, 81, 82, 83, 84, 85, 86, 88, 89, 90, 91, 92, 93, 94, 96, 97, 98, 99, 100, 102, 111, 144
  coastal, 1, 2, 3, 4, 6, 7, 8, 9, 10, 13, 15, 19, 29, 31, 34, 39, 40, 41, 48, 55, 58, 62, 63, 67, 71, 77, 79, 82, 84, 91, 92, 95, 97, 144
  endemic, xii, 7, 12, 13, 43, 44, 46, 47, 48
  hybrid, xiii, 2, 10, 13, 37, 77, 78
  invasive, 2, 11, 13, 51, 52, 53, 54, 76, 93, 98
  native, 4, 10, 11, 13, 53, 80, 95, 100
  naturalized, 13, 53
  xeromorphic, 13
  xerophilous, 2, 3
Spodiopogon sibiricus Trin., 42, 129
Stenofetuca pauciflora, 12
Stipa baicalensis, 66
strategy, 2, 9
  adaptive, vii, xi, 2, 3, 19, 21, 22, 25, 26, 33, 37, 73, 83
Syneilesis aconitifolia (Bunge) Maxim., 130

T

taxon (taxa)
  hybrid, xiii, 2, 10, 13, 37, 77, 78
Taxus cuspidata, 3
tetraploid, 50
Thermopsis lupinoides (L.) Link, 130
Thymus disjunctus, 66
Thymus ternejicus, 6, 7, 9
tolerance, 18, 21, 23, 34
Trifolium lupinaster L., 130
Trifolium pacificum Bobr., 130
Trifolium pratense L., 130
Trifolium repens L., 130
Tripleurospermum inodorum (L.) Sch. Bip., 130
Trisetum molle (Michx.) Trin., 46
Trisetum sibiricum Rupr., 42, 130
Trisetum spicatum (L.) K. Richt., 130
Trollius chinensis Bunge, 131
Truellum thunbergii (Siebold. et Zucc) Soják, 131
Turczaninowia fastigiata (Fisch.) DC., 131

U

Urtica angustifolia Fisch. ex Hornem., 131
Urtica platyphylla Wedd., 131

V

variability, xi, 2, 3, 10, 16, 19, 20, 33, 39, 43, 46, 53, 55, 57, 58, 64, 67, 73, 76, 111
  coenopopulation, 16, 20, 21, 49, 80
  karyological, xiii, 7, 15

vegetation, vii, xi, xii, 1, 2, 3, 15, 16, 17, 18, 19, 20, 21, 23, 32, 35, 45, 49, 50, 54, 63, 64, 71, 75, 77, 79, 80, 81, 82, 83, 85, 88, 92, 94, 95, 96, 97, 100, 101, 102, 144
  coastal, 1, 2, 3, 4, 6, 7, 8, 9, 10, 13, 15, 19, 29, 31, 34, 39, 40, 41, 48, 55, 58, 62, 63, 67, 71, 77, 79, 82, 84, 91, 92, 95, 97, 144
  desert, 2
Veratrum maackii Regel, 131
Vicia amoena, 68
Vicia amurensis, 66
Vicia cracca L., 131
Vicia unijuga A. Br., 131
Viola acuminata Ledeb., 131
Viola kamtschadalorum, 7
Viola patrinii Ging., 131

Z

Zizania latifolia (Griseb.) Stapf, 44, 131
zone, vii, xi, xii, xiii, 1, 2, 3, 5, 6, 7, 8, 9, 10, 13, 15, 16, 18, 19, 21, 34, 39, 41, 43, 51, 61, 64, 72, 73, 79, 87, 89, 91, 92, 95, 97, 98, 143, 145
  arid, 2, 3, 6
  climatic, 11, 12, 73
  contact, iv, xi, 1, 2, 5, 6, 8, 10, 15, 39, 70, 73, 89, 91, 92, 98, 143, 145
  hybrid, xiii, 2, 10, 13, 37, 77, 78
  land-sea, 1, 2, 15
  monsoon,, vii, xi, xii, xiii, 1, 3, 6, 8, 9, 13, 15, 16, 18, 19, 21, 34, 39, 41, 43, 50, 51, 58, 61, 64, 72, 73, 75, 76, 79, 83, 84, 85, 88, 90, 91, 93, 94, 97, 99, 103
  supralittoral, 1, 49, 50
  temperate, xi, xii, 81, 88, 103
Zostera japonica, 4, 7